Group Theory for Chemists

"Talking of education, people have now a-days" (said he) "got a strange opinion that every thing should be taught by lectures. Now, I cannot see that lectures can do so much good as reading the books from which the lectures are taken. I know nothing that can be best taught by lectures, except where experiments are to be shewn. You may teach chymestry by lectures — You might teach making of shoes by lectures!"

From James Boswell's *Life of Samuel Johnson*, 1766

About the Author

Kieran Molloy was born in Smethwick, England and educated at Halesowen Grammar School after which he studied at the University of Nottingham where he obtained his BSc, which was followed by a PhD degree in chemistry, specialising in main group organometallic chemistry. He then accepted a postdoctoral position at the University of Oklahoma where he worked in collaboration with the late Professor Jerry Zuckerman on aspects of structural organotin chemistry of relevance to the US Navy.

His first academic appointment was at the newly established National Institute for Higher Education in Dublin (now Dublin City University), where he lectured from 1980 to 1984. In 1984, Kieran Molloy took up a lectureship at the University of Bath, where he has now become Professor of Inorganic Chemistry. His many research interests span the fields of synthetic and structural inorganic chemistry with an emphasis on precursors for novel inorganic materials.

In 2003 he was joint recipient of the Mary Tasker prize for excellence in teaching, an award given annually by the University of Bath based on nominations by undergraduate students. This book *Group Theory for Chemists* is largely based on that award-winning lecture course.

Group Theory for Chemists

Fundamental Theory and Applications

Second Edition

Kieran C. Molloy

Oxford Cambridge Philadelphia New Delhi

Published by Woodhead Publishing Limited, 80 High Street, Sawston,
Cambridge CB22 3HJ, UK
www.woodheadpublishing.com

Woodhead Publishing, 1518 Walnut Street, Suite 1100, Philadelphia, PA 19102-3406, USA

Woodhead Publishing India Private Limited, 303, Vardaan House, 7/28 Ansari Road, Daryaganj, New Delhi – 110002, India
www.woodheadpublishingindia.com

First edition 2004, Horwood Publishing Limited
Second edition 2011, Woodhead Publishing Limited
Reprinted with corrections, 2013

British Library Cataloguing in Publication Data
A catalogue record for this book is available from the British Library.

ISBN 978-0-85709-240-3 (print)
ISBN 978-0-85709-241-0 (online)

The publisher's policy is to use permanent paper from mills that operate a sustainable forestry policy, and which has been manufactured from pulp which is processed using acid-free and elemental chlorine-free practices. Furthermore, the publisher ensures that the text paper and cover board used have met acceptable environmental accreditation standards.

Table of Contents

Preface ix

Part I
Symmetry and Groups

1. Symmetry

1.1 Symmetry 3
1.2 Point Groups 8
1.3 Chirality and Polarity 14
1.4 Summary 15
Problems 16

2. Groups and Representations

2.1 Groups 17
2.2 Transformation Matrices 19
2.3 Representations of Groups 20
2.4 Character Tables 25
2.5 Symmetry Labels 27
2.6 Summary 28
Problems 29

Part II
Application of Group Theory to Vibrational Spectroscopy

3. Reducible Representations

3.1 Reducible Representations 33
3.2 The Reduction Formula 37
3.3 The Vibrational Spectrum of SO_2 38
3.4 Chi Per Unshifted Atom 41
3.5 Summary 44
Problems 44

4. Techniques of Vibrational Spectroscopy

4.1 General Considerations 46
4.2 Infrared Spectroscopy 48
4.3 Raman Spectroscopy 49
4.4 Rule of Mutual Exclusion 50
4.5 Summary 53
Problems 53

5. The Vibrational Spectrum of $Xe(O)F_4$

5.1 Stretching and Bending Modes 55
5.2 The Vibrational Spectrum of $Xe(O)F_4$ 60
5.3 Group Frequencies 63
Problems 65

Part III
Application of Group Theory to Structure and Bonding

6. Fundamentals of Molecular Orbital Theory
6.1 Bonding in H_2 71
6.2 Bonding in Linear H_3 72
6.3 Limitations in a Qualitative Approach 75
6.4 Summary 77
Problems 77

7. H_2O – Linear or Angular ?
7.1 Symmetry-Adapted Linear Combinations 79
7.2 Central Atom Orbital Symmetries 80
7.3 A Molecular Orbital Diagram for H_2O 81
7.4 A C_{2v} / $D_{\infty h}$ MO Correlation Diagram 82
7.5 Summary 85
Problems 85

8. NH_3 – Planar or Pyramidal ?
8.1 Linear or Triangular H_3 ? 86
8.2 A Molecular Orbital Diagram for BH_3 89
8.3 Other Cyclic Arrays 91
8.4 Summary 95
Problems 95

9. Octahedral Complexes
9.1 SALCs for Octahedral Complexes 98
9.2 *d*-Orbital Symmetry Labels 100
9.3 Octahedral P-Block Complexes 101
9.4 Octahedral Transition Metal Complexes 102
9.5 π-Bonding and the Spectrochemical Series 103
9.6 Summary 105
Problems 106

10. Ferrocene
10.1 Central Atom Orbital Symmetries 110
10.2 SALCs for Cyclopentadienyl Anion 110
10.3 Molecular Orbitals for Ferrocene 113
Problems 116

Part IV
Application of Group Theory to Electronic Spectroscopy

11. Symmetry and Selection Rules
11.1 Symmetry of Electronic States 121
11.2 Selection Rules 123
11.3 The Importance of Spin 125

11.4 Degenerate Systems 126
11.5 Epilogue – Selection Rules for Vibrational Spectroscopy 130
11.6 Summary 131
Problems 131

12. Terms and Configurations
12.1 Term Symbols 134
12.2 The Effect of a Ligand Field – Orbitals 137
12.3 Symmetry Labels for d^n Configurations – An Opening 139
12.4 Weak Ligand Fields, Terms and Correlation Diagrams 142
12.5 Symmetry Labels for d^n Configurations – Conclusion 148
12.6 Summary 149
Problems 151

13. *d-d* Spectra
13.1 The Beer-Lambert Law 152
13.2 Selection Rules and Vibronic Coupling 153
13.3 The Spin Selection Rule 156
13.4 *d-d* Spectra – High-Spin Octahedral Complexes 157
13.5 *d-d* Spectra – Tetrahedral Complexes 160
13.6 *d-d* Spectra – Low-Spin Complexes 162
13.7 Descending Symmetry 164
13.8 Summary 169
Problems 170

Appendices

Appendix 1 Projection Operators 173
Appendix 2 Microstates and Term Symbols 182
Appendix 3 Answers to SAQs 185
Appendix 4 Answers to Problems 203
Appendix 5 Selected Character Tables 218

Index 223

Preface

The book I have written is based on a course of approximately 12 lectures and 6 hours of tutorials and workshops given at the University of Bath. The course deals with the basics of group theory and its application to the analysis of vibrational spectra and molecular orbital theory. As far as possible I have tried to further refer group theory to other themes within inorganic chemistry, such as the links between VSEPR and MO theory, crystal field theory (CFT) and electron deficient molecules. The book is aimed exclusively at an undergraduate group with a highly focussed content and thus topics such as applications to crystallography, electronic spectra etc have been omitted. The book is organised to parallel the sequence in which I present the material in my lectures and is essentially a text book which can be used by students as consolidation. However, group theory can only be mastered and appreciated by problem solving, and I stress the importance of the associated problem classes to my students. Thus, I have interspersed self-assessment questions to reinforce material at key stages in the book and have added additional exercises at the end of most chapters. In this sense, my offering is something of a hybrid of the books by Davidson, Walton and Vincent.

I have made two pragmatic decisions in preparing this book. Firstly, there is no point in writing a textbook that nobody uses and the current vogue among undergraduates is for shorter, more focussed texts that relate to a specific lecture course; longer, more exhaustive texts are likely to remain in the bookshop, ignored by price-conscious purchasers who want the essentials (is it on the exam paper ?) and little more. Secondly, the aim of a textbook is to inform and there seems to me little point in giving a heavily mathematical treatment to a generation of students for whom numbers are an instant turn-off. I have thus adopted a qualitative, more pictorial approach to the topic than many of my fellow academics might think reasonable. The book is thus open to the inevitable criticism of being less than rigorous, but, as long as I have not distorted scientific fact to the point of falsehood, I am happy to live with this.

Note for Students

Group theory is a subject that can only be mastered by practising its application. It is not a topic which lends itself to rote-learning, and requires an *understanding* of the methodology, not just a *knowledge* of facts. The self-assessment questions (*SAQ*s) which can be found throughout the book are there to test your understanding of the information which immediately precedes them. You are strongly advised to tackle each of these *SAQ*s as they occur and to check your progress by reference to the answers given in Appendix 3. Longer, more complex problems, some with answers, can be found at the end of each chapter and should be used for further consolidation of the techniques.

Note for Lecturers

In addition to the *SAQ*s and problems for which answers have been provided there are a number of questions at the end of most chapters for which solutions have not been given and which may be useful for additional tutorial or assessment work.

Acknowledgements

In producing the lecture course on which this book is based I relied heavily on the textbook *Group Theory for Chemists* by Davidson, perhaps naturally as that author had taught me the subject in my undergraduate years. Sadly, that book is no longer in print – if it were I would probably not have been tempted to write a book of my own. I would, however, like to acknowledge the influence that book *primus inter pares** has had on my approach to teaching the subject.

I would also like to offer sincere thanks to my colleagues at the University of Bath – Mary Mahon, Andy Burrows, Mike Whittlesey, Steve Parker, Paul Raithby – as well as David Cardin from the University of Reading, for their comments, criticisms and general improvement of my original texts. In particular, though, I would like to thank David Liptrot, a Bath undergraduate, for giving me a critical student view of the way the topics have been presented. Any errors and shortcomings that remain are, of course, entirely my responsibility.

Kieran Molloy
University of Bath, August 2010

* Other texts on chemical group theory, with an emphasis on more recent works, include:

J S Ogden, *Introduction to Molecular Symmetry (Oxford Chemistry Primers 97)*, OUP, 2001.

A Vincent, *Molecular Symmetry and Group Theory: A Programmed Introduction to Chemical Applications,* 2nd Edition, John Wiley and Sons, 2000.

P H Walton, *Beginning Group Theory for Chemistry*, OUP, 1998.

M Ladd, *Symmetry and Group Theory in Chemistry*, Horwood Chemical Science Series, 1998.

R. L. Carter, *Molecular Symmetry and Group Theory*, Wiley and Sons, 1998.

G Davidson, *Group Theory for Chemists*, Macmillan Physical Science Series, 1991.

F A Cotton, *Chemical Applications of Group Theory*, 3rd Edition, John Wiley and Sons, 1990.

PART I

SYMMETRY AND GROUPS

1

Symmetry

While everyone can appreciate the appearance of symmetry in an object, it is not so obvious how to classify it. The amide (**1**) is less symmetric than either ammonia or borane, but which of ammonia or borane – both clearly "symmetric" molecules – is the more symmetric ? In (**1**) the single N-H bond is clearly unique, but how do the three N-H bonds in ammonia behave ? Individually or as a group ? If as a group, how ? Does the different symmetry of borane mean that the three B-H bonds will behave differently from the three N-H bonds in ammonia ? Intuitively we would say "yes", but can these differences be predicted ?

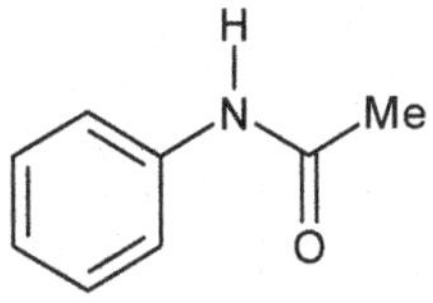

(1)

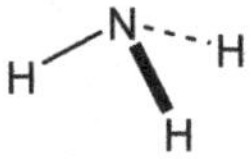

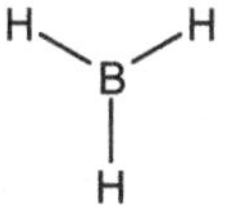

This opening chapter will describe ways in which the symmetry of a molecule can be classified (**symmetry elements** and **symmetry operations**) and also to introduce a shorthand notation which embraces all the symmetry inherent in a molecule (a **point group symbol**).

1.1 SYMMETRY

Imagine rotating an equilateral triangle about an axis running through its mid-point, by 120° (*overleaf*). The triangle that we now see is different from the original, but unless we label the corners of the triangle so we can follow their movement, it is *indistinguishable* from the original.

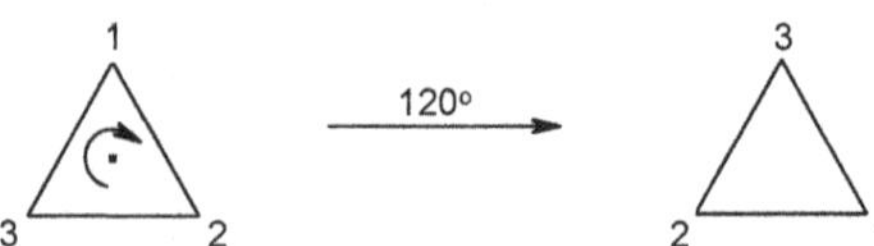

The symmetry inherent in an object allows it to be moved and still leave it looking unchanged. We define such movements as **symmetry operations,** e.g. a rotation, and each symmetry operation must be performed with respect to a **symmetry element**, which in this case is the rotation axis through the mid-point of the triangle.

It is these symmetry elements and symmetry operations which we will use to classify the symmetry of a molecule and there are four symmetry element / operation pairs that need to be recognised.

1.1.1 Rotations and Rotation Axes

In order to bring these ideas of symmetry into the molecular realm, we can replace the triangle by the molecule BF_3, which valence-shell electron-pair repulsion theory (VSEPR) correctly predicts has a trigonal planar shape; the fluorine atoms are labelled only so we can track their movement. If we rotate the molecule through 120° about an axis perpendicular to the plane of the molecule and passing through the boron, then, although the fluorine atoms have moved, the resulting molecule is indistinguishable from the original. We could equally rotate through 240°, while a rotation through 360° brings the molecule back to its starting position. Each of these rotations is a symmetry operation and the symmetry element is the rotation axis passing through boron.

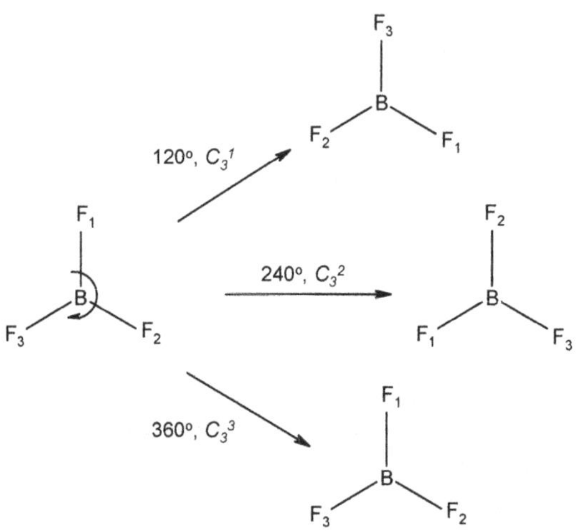

Fig. 1.1 Rotation as a symmetry operation.

Remember, all symmetry operations must be carried out with respect to a symmetry element. The symmetry element, in this case the rotation axis, is called a **three-fold axis** and is given the symbol C_3. The three operations, rotating about 120°, 240° or 360°, are given the symbols C_3^1, C_3^2 and C_3^3, respectively. The operations C_3^1 and C_3^2 leave the molecule indistinguishable from the original, while only C_3^3 leaves it identical. These two scenarios are, however, treated equally for identifying symmetry.

In general, an n-fold C_n axis generates n symmetry operations corresponding to rotations through multiples of $(360 / n)^\circ$, each of which leaves the resulting molecule indistinguishable from the original. A rotation through $m \times (360 / n)^\circ$ is given the symbol C_n^m. Table 1 lists the common rotation axes, along with examples.

Table 1.1 Examples of common rotation axes.

		C_n	rotation angle, °
H_2O	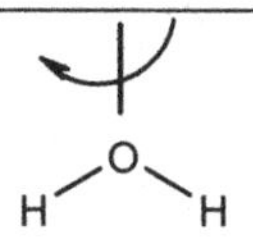	C_2	180
PCl_3	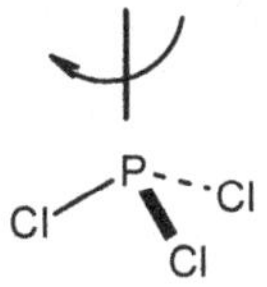	C_3	120
XeF_4	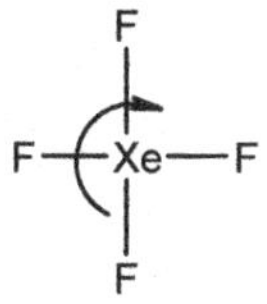	C_4	90
$[C_5H_5]^-$	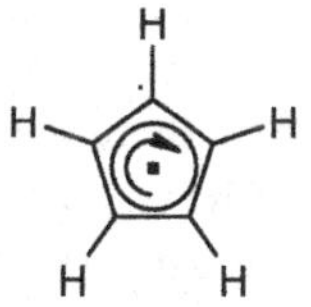	C_5	72
C_6H_6	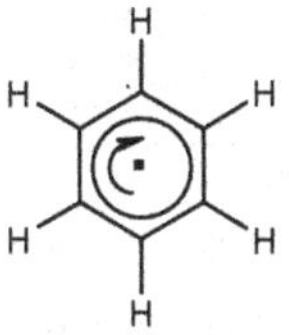	C_6	60

Where more than one rotation axis is present in a molecule, the one of **highest order** (maximum n) is called the **main** (or **principal**) **axis**. For example, while $[C_5H_5]^-$ also contains five C_2 axes (along each C-H bond), the C_5 axis is that of highest order. Furthermore, some rotations can be classified in more than one way. In benzene, C_6^3 is the same as C_2^1 about a C_2 axis coincident with C_6. Similarly, C_6^2 and C_6^4 can be classified as operations C_3^1 and C_3^2 performed with respect to a C_3 axis also coincident with C_6.

The operation C_n^n, e.g. C_2^2, always represents a rotation of 360° and is the equivalent of doing nothing to the object. This is called the **identity** operation and is given the symbol E (from the German *Einheit* meaning unity).

> *SAQ 1.1 : Identify the rotation axes present in the molecule cyclo-C_4H_4 (assume totally delocalised π-bonds). Which one is the principal axis ?*
>
> **Answers to all SAQs are given in Appendix 3.**

1.1.2 Reflections and Planes of Symmetry

The second important symmetry operation is **reflection** which takes place with respect to a **mirror plane**, both of which are given the symbol σ. Mirror planes are usually described with reference to the Cartesian axes *x*, *y*, *z*. For water, the *xz* plane is a mirror plane :

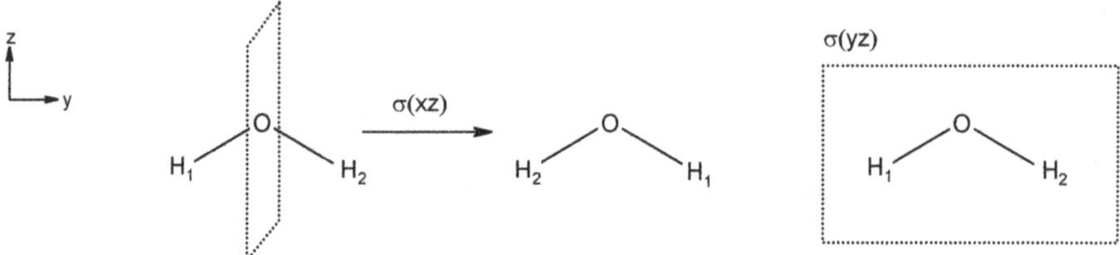

Fig. 1.2 Reflection as a symmetry operation.

Water has a second mirror plane, $\sigma(yz)$, with all three atoms lying in the mirror plane. Here, reflection leaves the molecule identical to the original.
Unlike rotation axes, mirror planes have only one associated symmetry operation, as performing two reflections with respect to the same mirror plane is equivalent to doing nothing i.e. the identity operation, *E*. However, there are three types of mirror plane that need to be distinguished. A **horizontal mirror plane** σ_h is one which is perpendicular to the main rotation axis. If the mirror plane contains the main rotation axis it is called a **vertical plane** and is given the symbol σ_v. Vertical planes which bisect bond angles (more strictly, one which bisects two σ_v or two C_2 operations) are called **dihedral planes** and labelled σ_d, though in practice σ_v and σ_d can be treated as being the same when assigning point groups (*Section 1.2.1*).

Square planar $[PtCl_4]^{2-}$ contains examples of all three types of mirror plane (Fig. 1.3); σ_h contains the plane of the molecule and is perpendicular to the main C_4 rotation axis, while σ_v and σ_d lie perpendicular to the molecular plane.

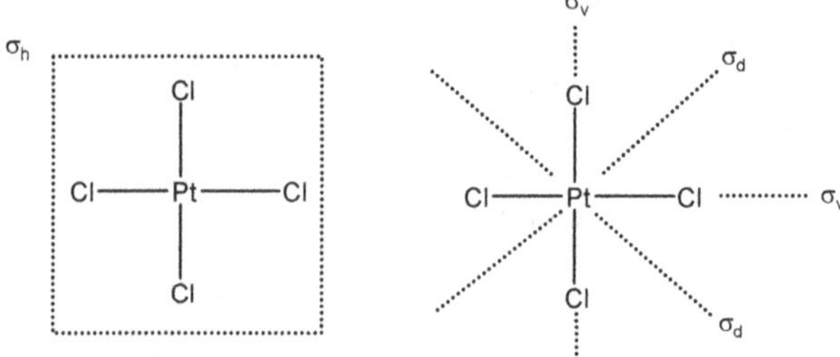

Fig 1.3 Examples of three different kinds of mirror plane.

SAQ 1.2 : Locate examples of each type of mirror plane in trans-$MoCl_2(CO)_4$.

1.1.3 Inversion and Centre of Inversion

The operation of **inversion** is carried out with respect to a **centre of inversion** (also referred to as a **centre of symmetry**) and involves moving every point (x, y, z) to the corresponding position $(-x, -y, -z)$. Both the symmetry element and the symmetry operation are given the symbol i. Molecules which contain an inversion centre are described as being **centrosymmetric**.

This operation is illustrated by the octahedral molecule SF_6, in which F_1 is related to F_1' by inversion through a centre of inversion coincident with the sulphur (Fig. 1.4a); F_2 and F_2', F_3 and F_3' are similarly related. The sulphur atom, lying on the inversion centre, is unmoved by the operation. Inversion centres do not *have* to lie on an atom. For example, benzene has an inversion centre at the middle of the aromatic ring.

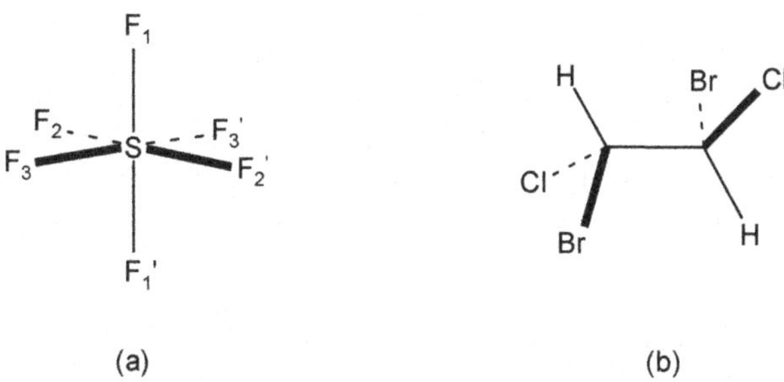

Fig. 1.4 Examples of inversion in which the inversion centre lies (a) on and (b) off an atomic centre.

Although inversion is a unique operation in its own right, it can be broken down into a combination of two separate operations, namely a C_2 rotation and a reflection σ_h. Rotation through 180° about an axis lying along z moves any point (x, y, z) to $(-x, -y, z)$. If this is followed by a reflection in the xy plane (σ_h because it is perpendicular to the z-axis) the point $(-x, -y, z)$ then moves to $(-x, -y, -z)$.

Any molecule which possesses both a C_2 axis and a σ at right angles to it as symmetry elements *must* also contain an inversion centre. However, the converse is not true, and it is possible for a molecule to possess i symmetry *without either of the other two symmetry elements being present.* The staggered conformation of the haloethane shown in Fig. 1.4b is such a case; here, the inversion centre lies at the mid-point of the C-C bond.

1.1.4 Improper Rotations and Improper Rotation Axes

The final symmetry operation/symmetry element pair can also be broken down into a combination of operations, as with inversion. An **improper rotation** (in contrast to a *proper rotation*, C_n) involves rotation about a C_n axis followed by a reflection in a mirror plane perpendicular to this axis (Fig. 1.5). This is the most complex of the symmetry operations and is easiest to understand with an example. If methane is rotated by 90° about a C_4 axis then reflected in a mirror plane perpendicular to this axis (σ_h), the result is indistinguishable from the original:

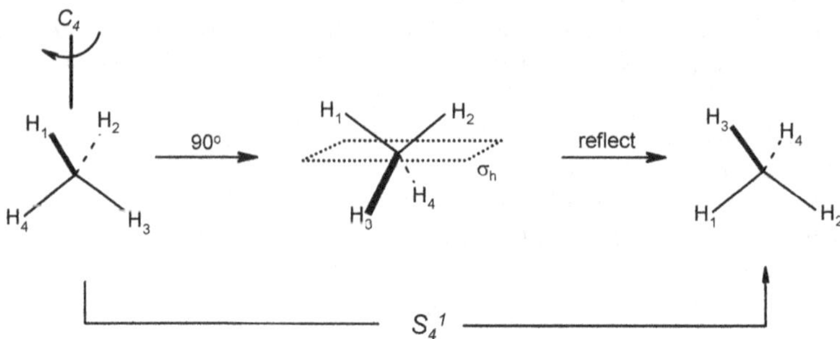

Fig 1.5 An improper rotation as the composite of a rotation followed by a reflection.

This complete symmetry operation, shown in Fig. 1.5, is called an **improper rotation** and is performed with respect to an **improper axis** (sometimes referred to as a **rotation-reflection axis** or **alternating axis**). The axis is given the symbol S_n (S_4 in the case of methane), where each rotation is by $(360\ /\ n)^\circ$ (90° for methane). Like a proper rotation axis, an improper axis generates several symmetry operations and which are given the notation S_n^m. When n is even, S_n^n is equivalent to E, e.g. in the case of methane, the symmetry operations associated with S_4 are S_4^1, S_4^2, S_4^3 and S_4^4, with $S_4^4 \equiv E$. It is important to remember that, for example, S_4^2 does *not* mean "rotate through 180° and then reflect in a perpendicular mirror plane". S_4^2 means "rotate through 90° and reflect" two consecutive times.

Note that an improper rotation is a unique operation, even though it can be thought of as combining two processes: methane does not possess either a C_4 axis or a σ_h mirror plane as individual symmetry elements but still possesses an S_4 axis. However, any molecule which *does* contain both C_n and σ_h *must* also contain an S_n axis. In this respect, improper rotations are like inversion.

SAQ 1.3 : Using IF_7 as an example, what value of m makes S_5^m equivalent to E ?

SAQ 1.4 : What symmetry operation is S_5^5 equivalent to ?

1.2 POINT GROUPS

While the symmetry of a molecule can be described by listing all the symmetry elements it possesses, this is cumbersome. More importantly, it is also unnecessary to locate all the symmetry elements since the presence of certain elements automatically requires the presence of others. You will have already noted (Section 1.1.3) that if a C_2 axis and σ_h can be identified, an inversion centre must also be present. As another example, in the case of BF_3 the principal axis (C_3) along with a σ_v lying along a B-F bond requires that two additional σ_v planes must also be present, generated by rotating one σ_v by either 120° or 240° about the main axis (Fig. 1.6).

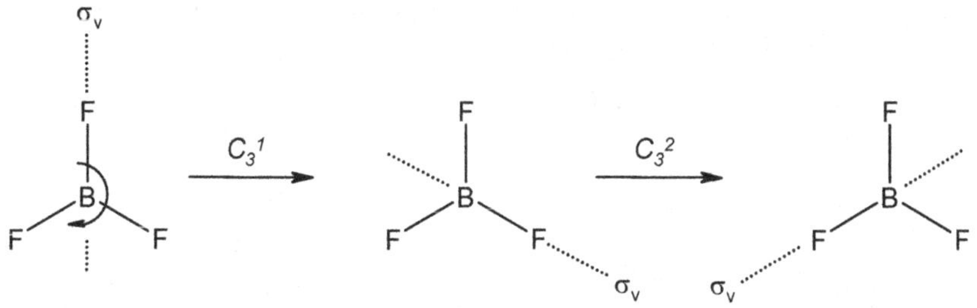

Fig. 1.6 Combination of C_3 and σ_v to generate two additional σ_v planes.

1.2.1 Point Group Classification

The symmetry elements for a molecule all pass through at least one point which is unmoved by these operations. We thus define a **point group** as a collection of symmetry elements (operations) and a **point group symbol** is a shorthand notation which identifies the point group. It is first of all necessary to describe the possible point groups which arise from various symmetry element combinations, starting with the lowest symmetry first. When this has been done, you will see how to derive the point group for a molecule without having to remember all the possibilities.

- the lowest symmetry point group has no symmetry other than a C_1 axis, i.e. E, and would be exemplified by the unsymmetrically substituted methane C(H)(F)(Cl)(Br). This is the C_1 point group.

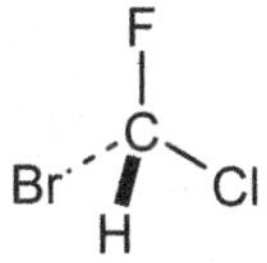

- molecules which contain only a mirror plane or only an inversion centre belong to the point groups C_s (e.g. $SO_2(F)Br$, Fig. 1.7) or C_i, (the haloethane, Fig. 1.4b), respectively.

- when only a C_n axis is present the point group is labelled C_n e.g. *trans*-1,3-difluorocyclopentane (Fig. 1.7).

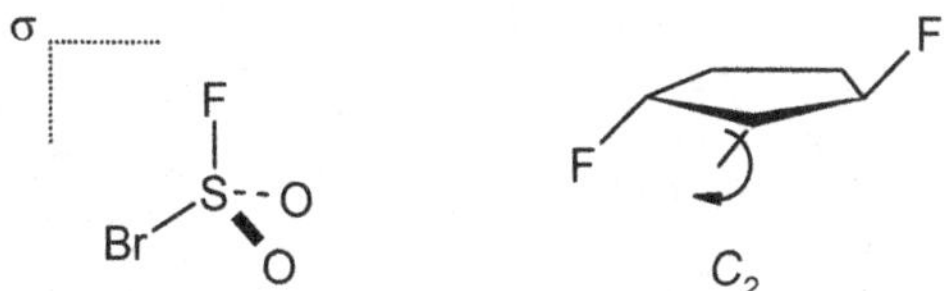

Fig. 1.7 Examples of molecules belonging to the C_s (*left*) and C_2 (*right*) point groups.

Higher symmetry point groups occur when a molecule possesses only one C_n axis but in combination with other symmetry elements (Fig. 1.8):

- a C_n axis in combination with n σ_v mirror planes gives rise to the C_{nv} family of point groups. H_2O (C_2 + 2 σ_v) is an example and belongs to the C_{2v} point group.

- where a C_n axis occurs along with a σ_h plane then the point group is C_{nh}. *Trans*-N_2F_2 (C_{2h}) is an example.
- molecules in which the C_n axis is coincident with an S_{2n} improper axis belong to the point group S_{2n} e.g. 1,3,5,7-F_4-cyclooctatetraene.

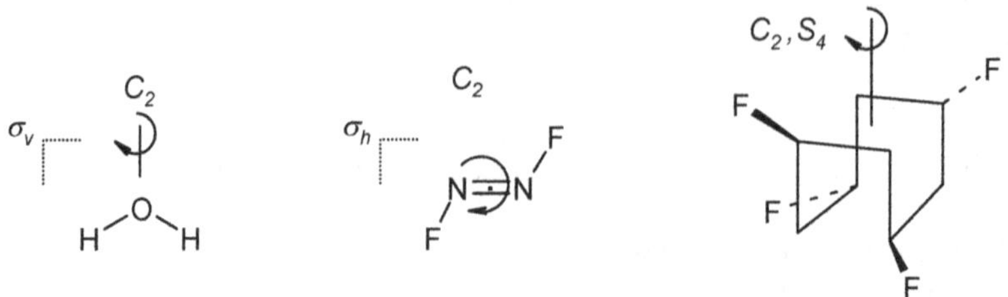

Fig. 1.8 Examples of molecules belonging to the C_{2v} (*left*), C_{2h} (*centre*) and S_4 (*right*) point groups. Double bonds of $C_8H_4F_4$ not shown for clarity.

SAQ 1.5 : What operation is S_2 equivalent to ? What point group is equivalent to the S_2 point group ?

Save for species of very high symmetry (*see below*), molecules which embody more than one rotation axis belong to families of point groups which begin with the designation D (so-called **dihedral point groups**). In these cases, in addition to a principal axis C_n the molecule will also possess n C_2 axes at right angles to this axis (Fig. 1.9). The presence of n C_2 axes is a consequence of the action of the C_n operations on one of the C_2 axes, in the same way that a C_n axis is always found in conjunction with n σ_v planes (Fig. 1.6) rather than just one.

- where no further symmetry elements are present the point group is D_n. An example here is the tris-chelate $[Co(en)_3]^{3+}$ (en = $H_2NCH_2CH_2NH_2$), shown schematically in Fig. 1.9 (*left*), which has a C_3 principal axis (direction of view) with three C_2 axes perpendicular (only one of which is shown).

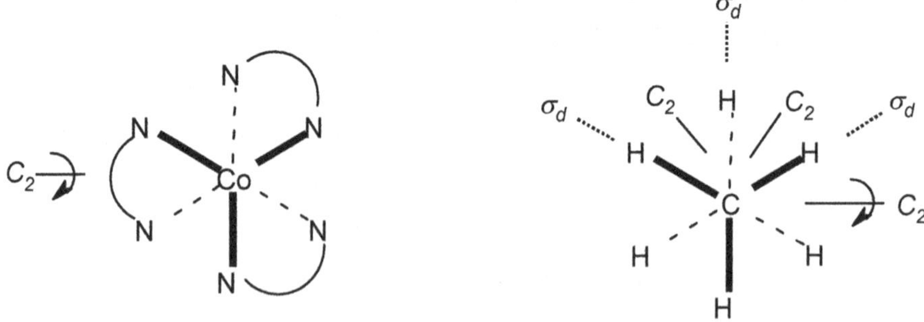

Fig. 1.9 Examples of molecules belonging to the D_3 (*left*) and D_{3d} (*right*) point groups.

- when C_n and n C_2 are combined with n vertical mirror planes the point group is D_{nd} The subscript *d* arises because the vertical planes (which by definition contain the C_n axis), bisect pairs of C_2 axes and are labelled σ_d. This point group family is exemplified by the staggered conformation of ethane, shown in Newman projection in Fig. 1.9 (*right*), viewed along the C-C bond.

- the D_{nh} point groups are common and occur when C_n and n C_2 are combined with a horizontal mirror plane σ_h; $[PtCl_4]^{2-}$ (Fig. 1.3) is an example (C_4, 4 C_2, σ_h).

Finally, there are a number of very high symmetry point groups which you will need to recognise. The first two apply to linear molecules, and are high symmetry versions of two of the point groups already mentioned (Fig. 1.10).

- $C_{\infty v}$ is the point group to which the substituted alkyne HC≡CF belongs. It contains a C_∞ axis along which the molecule lies (rotation about any angle leaves the molecule unchanged), in combination with ∞ σ_v mirror planes, one of which is shown in the figure.
- the more symmetrical ethyne HC≡CH belongs to the $D_{\infty h}$ point group. In addition to C_∞ and an infinite number of perpendicular C_2 axes (only one shown in the figure) the molecule also possesses a σ_h.

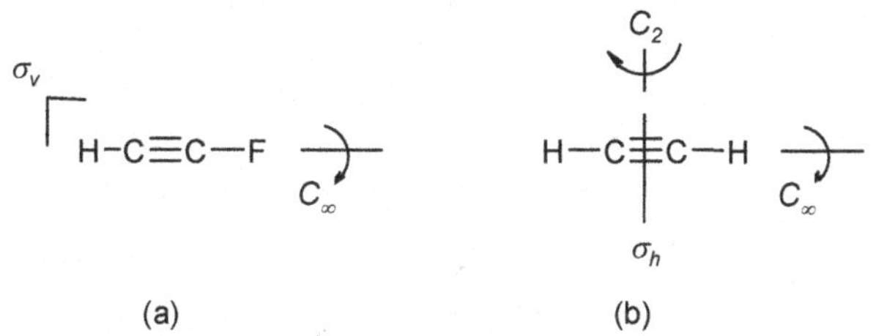

Fig. 1.10 Examples of (a) $C_{\infty v}$ and (b) $D_{\infty h}$ point groups.

In general, linear molecules which are centrosymmetric are $D_{\infty h}$ while non-centrosymmetric linear molecules belong to $C_{\infty v}$.

In addition to these systematically-named point groups, there are the so-called **cubic point groups** of which the two most important relate to perfectly tetrahedral (e.g. CH_4) and octahedral (e.g. SF_6) molecules. The two point groups which describe these situations are T_d and O_h, respectively. The term "cubic" arises because the symmetry elements associated with either shape relate to the symmetry of a cube. Moreover, visualisation of these elements is assisted by placing the molecule within the framework of a cube (*see Problem 1, below*). The 31 symmetry elements associated with O_h symmetry include C_4, C_3, C_2, S_4, S_2, σ_h, numerous vertical planes and i, some of which are shown in Fig. 1.11 for SF_6. Each face of the cube is identical, so C_2, C_4 and S_4 axes lie along each F-S-F unit, C_3 and S_6 axes pass through the centres of each of three pairs of triangular faces and C_2 axes pass through the centres of three opposite pairs of square faces. Another high symmetry arrangement, the perfect iscosahedron, which, though less common, is important in aspects of boron e.g. $[B_{12}H_{12}]^{2-}$ and materials chemistry e.g. C_{60}, has 120 symmetry elements and is given the symbol I_h.

It is worth emphasising at this point that the point group symbol T_d relates to symmetry and not a shape. The molecule $CHCl_3$ is tetrahedral in shape, but is not T_d

in symmetry (it is C_{3v}). Similarly, O_h refers to octahedral symmetry and not simply to an octahedral shape.

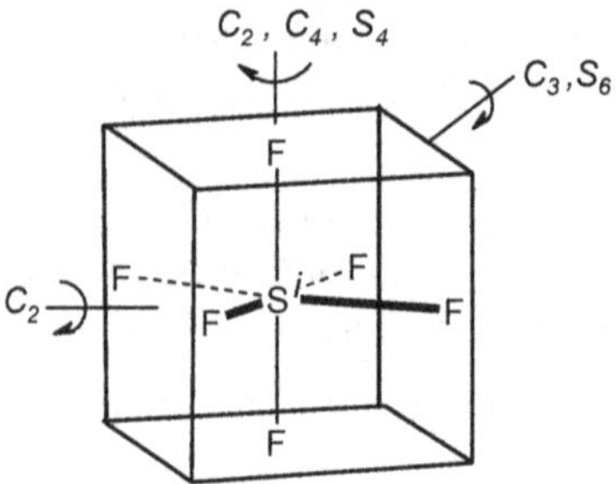

Fig. 1.11 Key symmetry elements of the O_h point group.

1.2.2 Assigning Point Groups

From the classification of point groups given above, it should be apparent that (*i*) not all symmetry elements need to be located in order to assign a point group and (*ii*) some symmetry elements take precedence over others. For example, $[PtCl_4]^{2-}$ has C_4 (and C_2 co-incident), four further C_2 axes perpendicular to C_4, σ_h, two σ_v, two σ_d, S_4 and i although only C_4, four C_2 axes perpendicular to C_4 and σ_h are required to classify the ion as D_{4h}. The hierarchy of symmetry elements occurs because, as you have already seen, certain combinations of symmetry elements automatically give rise to others. In this respect, σ_h takes precedence over σ_v and $[PtCl_4]^{2-}$ is D_{4h} and not D_{4d}.

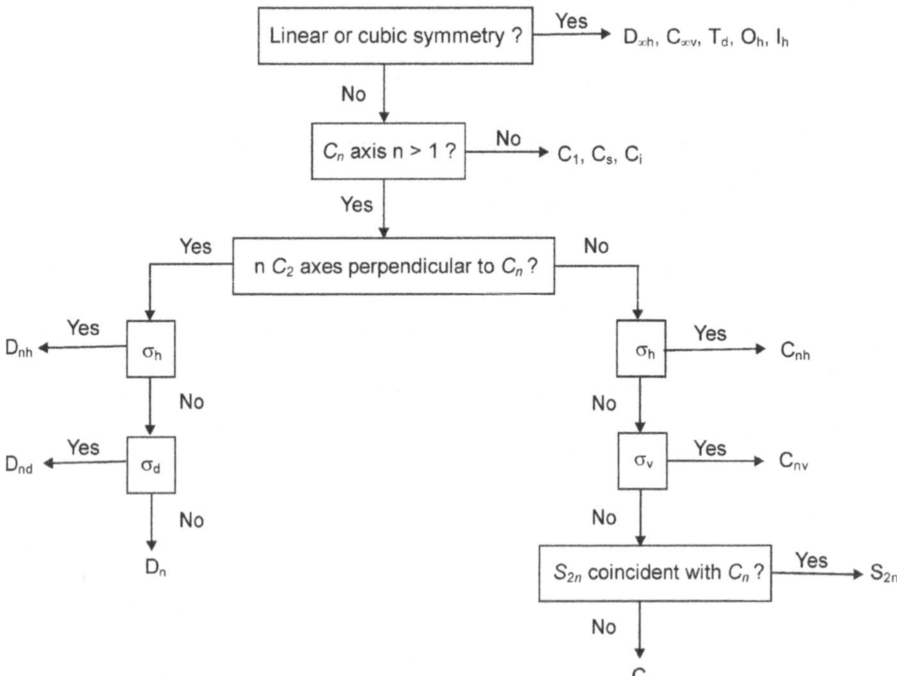

Fig.1.12 Flow chart showing the key decisions in point group assignment.

The flow chart shown in Fig. 1.12, along with the examples which follow, will highlight the key steps in assigning a molecule to its point group.

The sequence of questions that need to be asked, in order, are:

- does the molecule belong to one of the high symmetry (linear, cubic) point groups ?
- does the molecule possess a principal axis, C_n ?
- does the molecule possess n C_2 axes perpendicular to the principal axis C_n?
- does the molecule possess σ_h?
- does the molecule possess σ_v?

Example 1.1 : To which point group does PCl_3 belong ?

Firstly, the correct shape for PCl_3 (trigonal pyramidal) needs to be derived, in this case using VSEPR theory. The molecule clearly does not belong to one of the high symmetry point groups but does have a main axis, C_3 (Fig. 1.13). As there are no C_2 axes at right angles to C_3, the molecule belongs to a point group based on C_3 rather than D_3. While no σ_h is present, there are three σ_v (though it is only necessary to locate one of them), thus PCl_3 belongs to the C_{3v} point group.

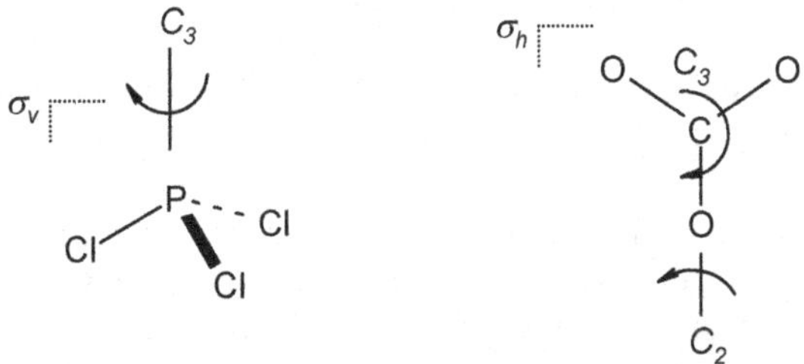

Fig. 1.13 Key symmetry elements in PCl_3 and $[CO_3]^{2-}$.

Example 1.2 : To which point group does $[CO_3]^{2-}$ belong ?

The carbonate anion is trigonal planar in shape. It is not linear or of high symmetry but it does possess a main axis, C_3. There are three C_2 axes perpendicular to this axis, so the point group is derived from D_3 (rather than C_3). The presence of σ_h makes the point group D_{3h}.

SAQ 1.6 : To which point group does PF_5 belong ?

Example 1.3 : To which point group does $S(O)Cl_2$ belong ?

$S(O)Cl_2$ is based on the tetrahedron but with one site occupied by a lone pair; the molecule is thus trigonal pyramidal in shape. The molecule is neither high symmetry nor does it possess an axis of symmetry higher than C_1. There is a mirror plane (containing S=O and bisecting the <Cl-S-Cl), so the point group is C_s.

Example 1.4 : To which point group does $[AsF_6]^-$ belong ?

The shape of $[AsF_6]^-$ is octahedral and, as each vertex of the octahedron is occupied by the same type of atom (F), the molecule has O_h symmetry.

1.3 CHIRALITY AND POLARITY

A chiral molecule is one which cannot be superimposed on its mirror image; each of the mirror images is termed an **enantiomer**. The most common example of chirality occurs when a molecule contains a carbon atom bonded to four different atoms (groups) (Fig. 1.14a); less easy to visualise are molecules which are chiral by virtue of their overall shape, such as "molecular propellers" in which three bidentate ligands chelate an octahedral metal centre e.g. $Cr(acac)_3$ (Fig. 1.14b):

(a) (b)

Fig. 1.14 Chiral molecules (a) with a chiral atom and (b) without a chiral atom; delocalised double bonds in the acac ligands have been omitted for clarity.

An alternative definition of chirality, given in terms of symmetry elements, is that:

- a non-linear molecule is chiral if it lacks an improper axis, S_n.

Note that the definition of an S_n axis includes both a mirror plane ($\sigma \equiv S_1$) and an inversion centre ($i \equiv S_2$), so chiral point groups are restricted to the C_n and D_n families which only require the presence of rotation axes. The definition of chirality in terms of symmetry elements can be particularly helpful in molecules where a well-defined chiral centre is absent, as in Fig. 1.14b.

It is important to appreciate that chiral molecules are not necessarily **asymmetric**, as this would imply that they have no symmetry at all. However, chiral molecules are **dissymmetric**, that is they may have some symmetry but lack an S_n axis. Asymmetric molecules *are* chiral, but only because they are dissymmetric molecules lacking any symmetry! The example in Fig. 1.14a has C_1 symmetry and is asymmetric, while the chromium chelate (Fig. 1.14b) is chiral but dissymmetric as it has both C_2 and C_3 symmetry elements (see Fig. 1.9a)

The absence or presence of a permanent dipole within a molecule is another key feature which has impact on, for example, spectroscopic properties. A dipole exists when the distribution of electrons within the molecule lacks certain symmetry, and, like chirality, can be defined in terms of symmetry elements and point groups. The clearest example of this is:

- any molecule which has an inversion centre i cannot have a permanent electric dipole.

This is because the electron density in one region is matched by the same electron density in the diametrically opposed region of the molecule, and thus no dipole is present. For similar reasons, other symmetry elements impose restrictions on the orientation of any dipole, but by themselves do not rule out its presence:

- a dipole cannot exist perpendicular to a mirror plane, σ.
- a dipole cannot exist perpendicular to a rotation axis, C_n.

It follows from this that certain combinations of symmetry elements also completely rule out the presence of a dipole. For example, any molecule possessing a C_n axis and either a C_2 or mirror plane perpendicular to this axis i.e. σ_h, cannot have a dipole. This means that molecules belonging to the following point groups are non-polar:

- any point group which includes an inversion centre, *i*.
- any D point group (D_n, D_{nh}, D_{nd}).
- any cubic point group (T_d, O_h, I_h).

SAQ 1.7 : Identify the point groups of the following species and hence state if they are (i) chiral and/or (ii) polar ?

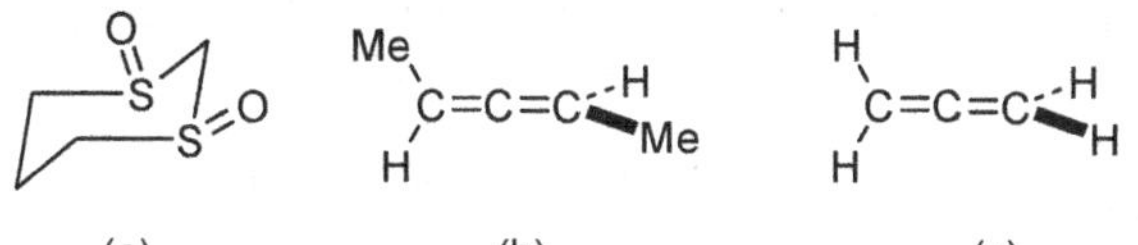

(a) (b) (c)

1.4 SUMMARY

- molecular symmetry can be classified in terms of symmetry operations, which are movements of the atoms which leave the molecule indistinguishable from the original.
- there are four symmetry operations: rotation (C_n), reflection (σ), inversion (i) and improper rotation (S_n).
- each symmetry operation is performed with respect to a symmetry element, which is either an axis (rotation), a plane (reflection), a point (inversion), or a combination of axis and plane perpendicular to this axis (an improper rotation).
- the axis of highest order is called the main, or principal, axis and has the highest value of n among the C_n axes present.
- mirror planes can be distinguished as σ_h (perpendicular to main axis), σ_v (containing the main axis) or σ_d (containing the main axis and bisecting bond angles), though σ_v and σ_d can be grouped together.
- rotations and improper axes can generate several operations (C_n^m, S_n^m) while only one operation is associated with either i or σ.
- a point group is a collection of symmetry elements (operations) and is identified by a point group symbol.
- a point group can be derived without identifying every symmetry element that is present, using the hierarchy outlined in the flow chart given in Fig. 1.12.
- molecules that do not possess an S_n axis are chiral.
- molecules that possess an inversion centre, or belong to either D or cubic point groups, are non-polar.

PROBLEMS

Answers to problems marked with * are given in Appendix 4.

1*. Identify the symmetry elements present in each of the following molecules:

NH_3 AsH_5 *cyclo*-$B_3N_3H_6$ (borazine) B(H)(F)(Br) SiH_4

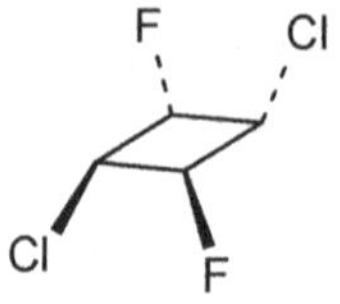

(*Hint*: *for* SiH_4, *place the tetrahedron inside a cube with alternating vertices occupied by hydrogen atoms; all faces of the cube are equivalent*).

2*. Identify the point groups of each of the following pairs of molecules:

CO_2 and SO_2
ferrocene (staggered) and ruthenocene (eclipsed)
cis- and *trans*-$Mo(CO)_4Cl_2$
$[IF_6]^+$ and $[IF_6]^-$
SnCl(F) and XeCl(F)
mer- and *fac*-WCl_3F_3

3. Link the following species with the correct point group:

$[AuCl_4]^-$	$[ClF_2]^+$	BrF_5	SO_3	$[ClF_2]^-$	$Ni(CO)_4$	$B(OH)_3$
C_{2v}	T_d	D_{3h}	$D_{\infty h}$	C_{3h}	C_{4v}	D_{4h}

4*. Which of the following species are chiral ? Which are polar ?

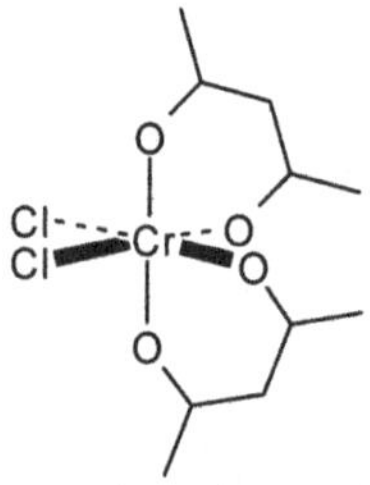

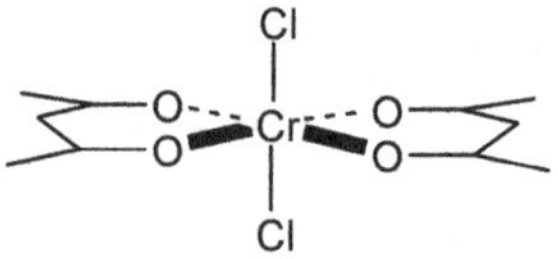

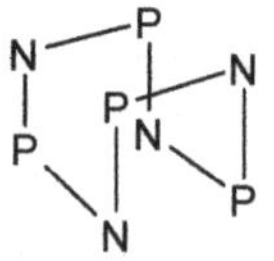

cyclo-$(Cl_2PN)_4$

In the schematic representation of the $Cr(acac)_2Cl_2$ delocalised double bonds within the acac ligands have been omitted for clarity. Similarly, for *cyclo*-$(Cl_2PN)_4$ the chlorine atoms and P=N double bonds have also been omitted.

2

Groups and Representations

While Chapter 1 outlined the concept of symmetry in a descriptive manner, this chapter will aim to place the concept on a more quantitative basis. Transformation matrices – the numerical descriptors of individual symmetry operations – will be introduced, as will numerical representations for the point group symmetry operations. These will allow group theory to be applied in a more quantitative way to the analysis of vibrational spectra and molecular orbital theory (*Parts II* and *III*).

2.1 GROUPS

A "group" can be thought of as an exclusive club to which only a number of members belong. These members must, in turn, agree to abide by certain rules. The collection of symmetry operations which make up a point group form such a group e.g. for C_{2v} the group members are E ($\equiv C_2^2$), C_2 (i.e. C_2^1) and two vertical planes $\sigma(xy)$, $\sigma(yz)$. Groups can be formed in many different ways, of which a collection of symmetry operations is only one. There are four mathematical rules which any group must obey, and are:

- the group must be closed i.e. any combination of two or more members of the group must be equivalent to another member of the group.

For example:

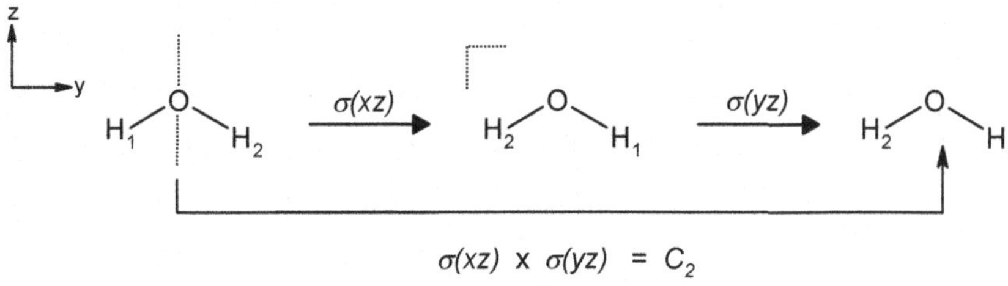

z
y
H_1 O H_2 $\xrightarrow{\sigma(xz)}$ H_2 O H_1 $\xrightarrow{\sigma(yz)}$ H_2 O H_1

$$\sigma(xz) \times \sigma(yz) = C_2$$

In this context the multiplication sign (×) should be read as "followed by".

SAQ 2.1 : Complete the following table and hence show that all operations of the C_{2v} point group form a closed group.

	E	C_2	$\sigma(xz)$	$\sigma(yz)$
E				
C_2				
$\sigma(xz)$				C_2
$\sigma(yz)$				

Entries in the Table correspond to operation 1(column) × operation 2(row). Note that some symmetry operations are equivalent, so that each box may have more than one entry.

Answers to all SAQs are given in Appendix 3.

- any group must contain one member such that it combines with any other member to leave it unchanged.

This is the identity element, E. i.e. $C_2 \times E = C_2$.

- every element must have an inverse, with which it combines to generate E.

For H_2O, C_2 is its own inverse i.e. $C_2 \times C_2 = E$.

SAQ 2.2 : What operation is the inverse of C_3^1 ? S_5^3 ?

- multiplication of members must be associative i.e. A × (B × C) = (A × B) × C.

e.g. $[\sigma(xz) \times \sigma(yz)] \times C_2$ should give the same product as $\sigma(xz) \times [\sigma(yz) \times C_2]$.

Since $[\sigma(xz) \times \sigma(yz)] = C_2$, (*see above*) then $[\sigma(xz) \times \sigma(yz)] \times C_2 = C_2 \times C_2 = E$.

Similarly, $[\sigma(yz) \times C_2] = \sigma(xz)$:

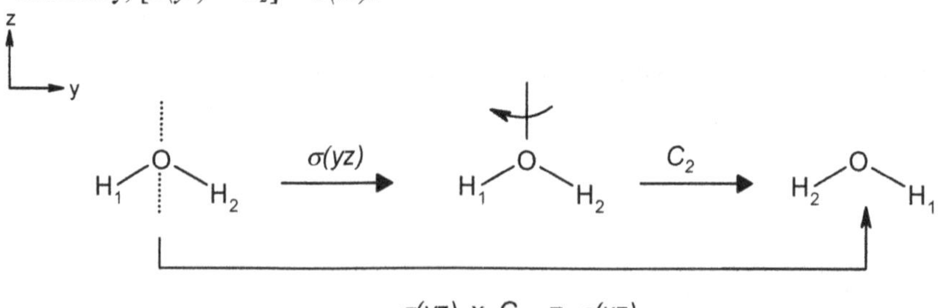

Therefore: $\sigma(xz) \times [\sigma(yz) \times C_2] = \sigma(xz) \times \sigma(xz) = E$.

2.2 TRANSFORMATION MATRICES

The mathematical tool for describing the effect of a symmetry operation is a **transformation matrix**. Before explaining how these matrices work, it is necessary to give a brief explanation of matrix multiplication.

A matrix is a grid of numbers with x rows and y columns, in which the position of a number is defined by its row and column. Matrices can be either square ($x = y$) or rectangular ($x \neq y$). In the following 2 × 2 ("two by two") matrix, the subscripts associated with each number refer to rows and columns:

$$\begin{bmatrix} x_{11} & x_{12} \\ x_{21} & x_{22} \end{bmatrix} \qquad x_{11} = \text{row 1, column 1};\ x_{12} = \text{row 1, column 2, etc}$$

Matrix "multiplication" involves combining a row of matrix 1 with a column of matrix 2 to generate a single entry in the product matrix, 3. Thus, the number of columns in matrix 1 must be the same as the number of rows of matrix 2. When the first row is combined with the first column the product is z_{11}. The following example shows how the entries z_{11}, z_{12} and z_{13} are derived:

$$\begin{bmatrix} x_{11} & x_{12} \\ x_{21} & x_{22} \\ x_{31} & x_{32} \end{bmatrix} \begin{bmatrix} y_{11} & y_{12} & y_{13} \\ y_{21} & y_{22} & y_{23} \end{bmatrix} = \begin{bmatrix} z_{11} & z_{12} & z_{13} \\ z_{21} & z_{22} & z_{23} \\ z_{31} & z_{32} & z_{33} \end{bmatrix}$$

$$z_{11} = x_{11}y_{11} + x_{12}y_{21} \qquad z_{12} = x_{11}y_{12} + x_{12}y_{22} \qquad z_{13} = x_{11}y_{13} + x_{12}y_{23}$$

Example 2.1 : What is the product of the following matrix multiplication ?

$$\begin{bmatrix} 1 & 4 \\ 3 & 2 \\ -1 & 0 \end{bmatrix} \begin{bmatrix} 2 & -2 & 0 \\ 1 & 1 & 3 \end{bmatrix} = \begin{bmatrix} 6 & 2 & 12 \\ 8 & -4 & 6 \\ -2 & 2 & 0 \end{bmatrix}$$

The entry z_{12} in the product matrix (row 1, column 2) is arrived at by multiplying the first row by the second column:

$$(1 \times -2) + (4 \times 1) = 2.$$

The product matrix (3 × 3) has the same number of rows as matrix 1 (3) and the same number of columns as matrix 2 (3).

SAQ 2.3 : What is the product of the following matrix multiplication?

$$\begin{bmatrix} 2 & 4 & -1 \\ 3 & 5 & 0 \\ -2 & 7 & 3 \end{bmatrix} \begin{bmatrix} 2 \\ 3 \\ -2 \end{bmatrix} =$$

Matrix multiplication can be used to denote, in a numerical manner, the transformation of a molecule brought about by the application of a symmetry operation. The first step is to identify the location to which each atom moves under a given symmetry operation and then construct the matrix which brings about this movement. Example 2.2 illustrates this.

Example 2.2 : Write a transformation matrix to describe the effect of the C_2 operation on the positions of the atoms of SO_2 (C_{2v}).

$$O^1 - S - O^2 \xrightarrow{C_2} O^2 - S - O^1$$

$$S \rightarrow S$$
$$O^1 \rightarrow O^2$$
$$O^2 \rightarrow O^1$$

$$\begin{bmatrix} 1 & 0 & 0 \\ 0 & 0 & 1 \\ 0 & 1 & 0 \end{bmatrix} \begin{bmatrix} S \\ O^1 \\ O^2 \end{bmatrix} = \begin{bmatrix} S \\ O^2 \\ O^1 \end{bmatrix}$$

SAQ 2.4 : What is the 5 × 5 matrix which describes the movement of the atoms of CH_4 under the operation S_4^1 (Fig. 1.5) ?

2.3 REPRESENTATIONS OF GROUPS

We have gone some way towards making symmetry transformations easier to describe accurately by replacing diagrammatic descriptions with matrix-based equations e.g. Example 2.2, *above*. However, we are still using symmetry descriptors such as C_3^1 and σ_h to describe the members of the point group. The next step is to replace these descriptors by numbers, which act as a **representation** of the point group. The numbers we chose to represent the symmetry operations of a group must themselves form an equivalent group i.e. multiplying two operations to generate a third requires that multiplying the two numbers that represent these operations must generate the number which represents the product operation. The other three rules required for the numbers (representations) to form a group i.e. the need for an identity (1), that each number have an inverse with which it combines to generate the identity and association, must also be obeyed.

One way to generate a point group representation is to look at the effect of the group symmetry operations on a series of vectors which describe the translational and rotational movement of atoms about the three Cartesian axes. These translational and rotational vectors are called **basis sets** as they are the basis on which the representations are derived. Atomic orbitals can also serve as basis sets and a final example will illustrate this. To begin with, however, we will focus on the use of vectors.

To illustrate the derivation of numerical representations of symmetry operations, consider the effect of the symmetry operations of the C_{2v} point group on the translation of H_2O along y and its rotation about z.

The three bold arrows (vectors) together describe the translation of the whole H_2O molecule along the y axis ($\mathbf{T_y}$). After the operation C_2 the vectors describe a translation in the $-y$ direction ($-\mathbf{T_y}$) so we can write the following equation to describe the action of the C_2 operation:

$$(C_2)\,\mathbf{T_y} = (-1)\,(\mathbf{T_y})$$

By a similar logic:

$$\begin{aligned}(E)\,\mathbf{T_y} &= (1)\,(\mathbf{T_y})\\ (\sigma(xz))\,\mathbf{T_y} &= (-1)\,(\mathbf{T_y})\\ (\sigma(yz))\,\mathbf{T_y} &= (1)\,(\mathbf{T_y})\end{aligned}$$

Thus, the integers 1, −1, −1 and 1 are a *representation* of the symmetry operations E, C_2, $\sigma(xz)$ and $\sigma(yz)$ that make up the C_{2v} point group.

This set of numbers is not unique, and other representations are also possible and can be found using any of the vectors describing $\mathbf{T_x}$, $\mathbf{T_z}$, $\mathbf{R_x}$, $\mathbf{R_y}$ or $\mathbf{R_z}$ as the basis set. For example, we can describe $\mathbf{R_z}$ by the following two vectors which "pull" one hydrogen and "push" the other to generate a rotation about the z axis:

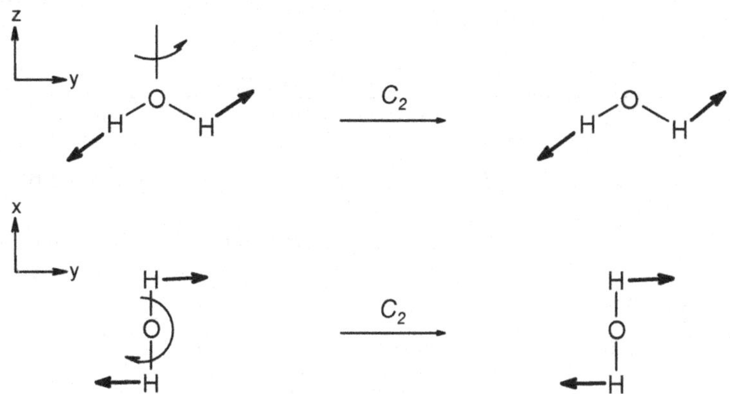

We can thus write:

$$\begin{aligned}(E)\,\mathbf{R_z} &= (1)\,(\mathbf{R_z})\\ (C_2)\,\mathbf{R_z} &= (1)\,(\mathbf{R_z})\\ (\sigma(xz))\,\mathbf{R_z} &= (-1)\,(\mathbf{R_z})\\ (\sigma(yz))\,\mathbf{R_z} &= (-1)\,(\mathbf{R_z})\end{aligned}$$

nd hence generate the integers 1, 1, −1 and −1 to represent the symmetry operations.

SAQ 2.5 : Show that the basis set of vectors describing $\boldsymbol{T_x}$ *generates a different representation for the operations of the* C_{2v} *point group from the representations generated by* $\boldsymbol{T_y}$ *and* $\boldsymbol{R_z}$*, above.*

It is not just the translational or rotational vectors which can be used as a basis to generate a representation of the point group - atomic orbitals can also be used. For example, using the p_y orbital as basis:

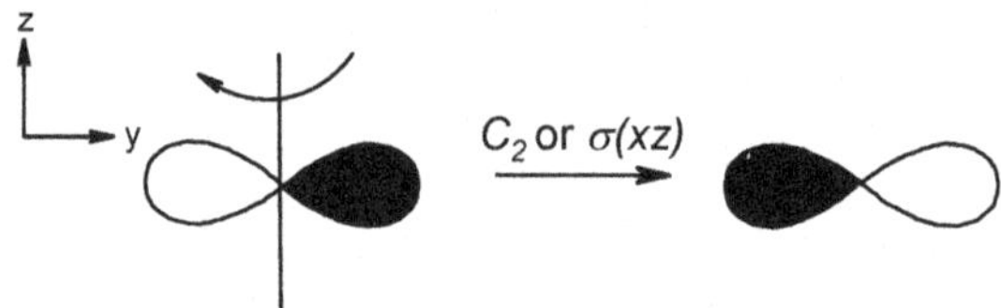

The different coloured lobes of the *p*-orbital represent positive and negative phases for the amplitude of the electron wave. We use −1 as the representation when the phase in any part of the orbital is reversed under a symmetry operation and 1 when it is unchanged. Thus, using these criteria and applying them to the four symmetry operations of C_{2v} results in the following:

E	C_2	$\sigma(xz)$	$\sigma(yz)$	
1	−1	−1	1	p_y

The representation arrived at this way is the same as that generated by using $\mathbf{T_y}$ (or $\mathbf{R_x}$) as the basis.

There are, in fact, only four separate representations of the C_{2v} point group that can be generated using either vectors or orbitals as basis sets, and these are summarised in Table 2.1

Table 2.1 Representations of the C_{2v} point group.

E	C_2	$\sigma(xz)$	$\sigma(yz)$	
1	1	1	1	$\mathbf{T_z}$
1	1	−1	−1	$\mathbf{R_z}$
1	−1	1	−1	$\mathbf{T_x}$, $\mathbf{R_y}$
1	−1	−1	1	$\mathbf{T_y}$, $\mathbf{R_x}$

Note that any of the four representations constitutes a group. For example, we have seen that $\sigma(xz) \times \sigma(yz) = C_2$ and that this is mimicked by the entries in the table – multiply any of the pairs of numbers in a row representing each of the two reflections and you will generate the number representing C_2 in that row. Each row contains an identity operation ($1 \equiv E$) and each representation has an inverse e.g. $C_2 \times C_2 = E$, and this is also reproduced by the entries under C_2 in the table.

> *SAQ 2.6 : Show that [$\sigma(xz) \times \sigma(yz)] \times C_2 = \sigma(xz) \times [\sigma(yz) \times C_2]$ using the representation of the C_{2v} point group generated by **T_x** as basis set.*

The four representations of the C_{2v} point group generated by the translations and rotations about the Cartesian axes are the only groups of simple integers which act this way, excluding the trivial representation 0, 0, 0, 0 which loses all the information about the way symmetry operations combine.

> *SAQ 2.7 : Show that the integers 1, −1, −1, −1 do not act as a representation of the C_{2v} point group.*

The representations of the C_{2v} point group shown in Table 2.1 are the simplest sets of integers that act as representations. Hence, they are termed **irreducible representations**. In general, the translational and rotational vectors are insufficient to generate all the irreducible representations for a group. However, all the irreducible representations of each point group have been derived and such information is readily available in tables known as **character tables**. Some character tables will be introduced below and further commonly used character tables are collected in Appendix 5.

While integers suffice to generate irreducible representations for the C_{2v} point group, this is not the case for all such groups. In the case of C_{2v} the translation / rotation vectors either transform onto themselves or their reverse under each of the symmetry operations; in general this is not true. Consider the case of NH_3 (C_{3v}):

Fig. 2.1 Vectors representing $\mathbf{R_z}$, $\mathbf{T_x}$,$\mathbf{T_y}$ and $\mathbf{T_z}$ for the C_{3v} point group. For clarity, the translational vectors on the hydrogens have been omitted.

The translation along, and rotation about, z ($\mathbf{T_z}$, $\mathbf{R_z}$) generate irreducible representations, as follows:

E	C_3^1	C_3^2	$\sigma_v(1)$	$\sigma_v(2)$	$\sigma_v(3)$	
1	1	1	1	1	1	$\mathbf{T_z}$
1	1	1	−1	−1	−1	$\mathbf{R_z}$

However, the vectors for $\mathbf{T_x}$ and $\mathbf{T_y}$ (Fig. 2.1 focuses solely on the vectors on nitrogen for simplicity) move to completely new positions under, for example C_3^1.

In this case, $\mathbf{T_x}$ and $\mathbf{T_y}$ need to be treated as a pair and the representation they generate is now not a simple integer but a matrix.

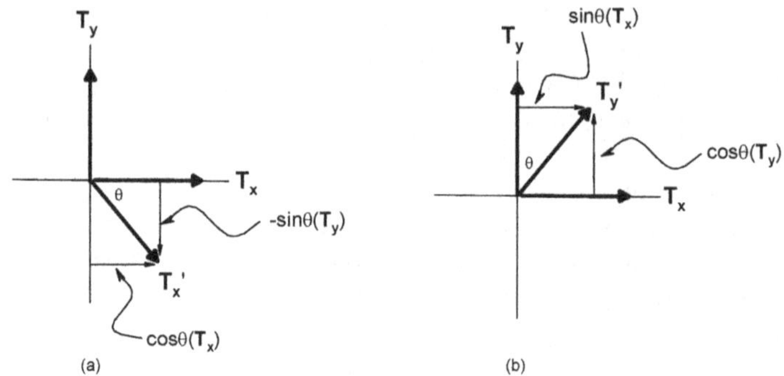

Fig 2.2 Rotation of either $\mathbf{T_x}$ or $\mathbf{T_y}$ about z generates new vectors $\mathbf{T_x}'$ and $\mathbf{T_y}'$ which can be seen as combinations of components of the original $\mathbf{T_x}$, $\mathbf{T_y}$ vectors.

When rotated through θ° (120° in the case of C_3), the vector $\mathbf{T_x}$ moves to a new position which can be described by combining components of the original vectors $\mathbf{T_x}$, $\mathbf{T_y}$ (Fig. 2. 2a):

$$\mathbf{T_x}' = \cos\theta\,(\mathbf{T_x}) \;-\; \sin\theta\,(\mathbf{T_y})$$

The "minus" in the $-\sin\theta\,(\mathbf{T_y})$ term is because the vector is in the opposite direction to that of $\mathbf{T_y}$. Similarly, rotating the vector $\mathbf{T_y}$ through θ° about z generates $\mathbf{T_y}'$, which can also be seen to be made up of components of $\mathbf{T_x}$ and $\mathbf{T_y}$ (Fig. 2.2b):

$$\mathbf{T_y}' = \sin\theta\,(\mathbf{T_x}) \;+\; \cos\theta\,(\mathbf{T_y})$$

The movement of $\mathbf{T_x}$ and $\mathbf{T_y}$ onto $\mathbf{T_x}'$ and $\mathbf{T_y}'$ can be written in terms of a transformation matrix:

$$\begin{bmatrix} \cos\theta & -\sin\theta \\ \sin\theta & \cos\theta \end{bmatrix} \begin{bmatrix} T_x \\ T_y \end{bmatrix} = \begin{bmatrix} T_x' \\ T_y' \end{bmatrix}$$

In the specific case of NH_3 and a C_3 rotation, the transformation matrix becomes:

$$\begin{bmatrix} -\tfrac{1}{2} & -\tfrac{\sqrt{3}}{2} \\ \tfrac{\sqrt{3}}{2} & -\tfrac{1}{2} \end{bmatrix} \begin{bmatrix} T_x \\ T_y \end{bmatrix} = \begin{bmatrix} T_x' \\ T_y' \end{bmatrix}$$

$$(\cos 120 = -1/2; \quad \sin 120 = \sqrt{3}/2 = 0.866 \text{ etc})$$

Similar transformation matrices can be derived for the $(\mathbf{T_x},\mathbf{T_y})$ pair of vectors for all the other operations of the C_{3v} point group, as can matrices for the pair of rotations $(\mathbf{R_x}, \mathbf{R_y})$. The complete table of irreducible representations for the C_{3v} point group is:

Table 2.2 Representations of the point group C_{3v}.

E	C_3^1	C_3^2	$\sigma_v(1)$	$\sigma_v(2)$	$\sigma_v(3)$	
1	1	1	1	1	1	$\mathbf{T_z}$
1	1	1	−1	−1	−1	$\mathbf{R_z}$
$\begin{bmatrix}1 & 0\\0 & 1\end{bmatrix}$	$\begin{bmatrix}-1/2 & -\sqrt{3}/2\\\sqrt{3}/2 & -1/2\end{bmatrix}$	$\begin{bmatrix}-1/2 & \sqrt{3}/2\\-\sqrt{3}/2 & -1/2\end{bmatrix}$	$\begin{bmatrix}1 & 0\\0 & -1\end{bmatrix}$	$\begin{bmatrix}-1/2 & -\sqrt{3}/2\\-\sqrt{3}/2 & 1/2\end{bmatrix}$	$\begin{bmatrix}-1/2 & \sqrt{3}/2\\\sqrt{3}/2 & 1/2\end{bmatrix}$	$\mathbf{(T_x, T_y)}$, $\mathbf{(R_x, R_y)}$

In fact, all the entries in Tables 2.1 and 2.2 are transformation matrices, with the integers being 1 × 1 matrices. As with the table of representations for C_{2v} (Table 2.1), each row of entries in Table 2.2 also forms a group.

SAQ 2.8 : For each of the three representations, show that multiplying the matrices that represent C_3^1 and C_3^2 generates the appropriate product matrices.

2.4 CHARACTER TABLES

We have seen in Section 2.3 that the symmetry operations of a point group can be replaced by a series of representations, each of which has a transformation matrix to do the job of the corresponding symmetry operation (e.g. Table 2.2). These matrices can be quite complex, but luckily, for reasons which are beyond the scope of this book, all the information required from the matrix is contained on the diagonal which runs from top left to bottom right.

- the sum of the numbers which lie on this "leading diagonal" is called the **character** of the matrix and is given the symbol χ ("chi")

In mathematical terms, this is written as

$$\chi = \sum_1^n z_{nn} = z_{11} + z_{22} + z_{33} \ldots\ldots\ldots + z_{nn}$$

but an example will illustrate the idea more easily.

Example 2.3 : What is the character of the following matrix:

$$\begin{bmatrix}1 & 2 & 6\\0 & -5 & 4\\3 & 2 & 8\end{bmatrix} \qquad \chi = 1 + (-5) + 8 = 4$$

*SAQ 2.9 : Derive the characters of the transformation matrices which represent the symmetry operations of the C_{3v} point group generated using **(T_x, T_y)**, **(R_x, R_y)**.*

Thus, the table of irreducible representations for C_{3v} can be re-written as:

C_{3v}	E	C_3^1	C_3^2	σ_v	σ_v	σ_v	
	1	1	1	1	1	1	$\mathbf{T_z}$
	1	1	1	−1	−1	−1	$\mathbf{R_z}$
	2	−1	−1	0	0	0	$(\mathbf{T_x},\mathbf{T_y})$ $(\mathbf{R_x}, \mathbf{R_y})$

Note that the representations of C_3^1 and C_3^2 are the same for the same basis vector, as are the representations for each of the three vertical planes. The table can therefore be simplified further as:

Table 2.3 The C_{3v} character table.

C_{3v}	E	$2C_3$	$3\sigma_v$		
A_1	1	1	1	$\mathbf{T_z}$	$x^2 + y^2, z^2$
A_2	1	1	−1	$\mathbf{R_z}$	
E	2	−1	0	$(\mathbf{T_x}, \mathbf{T_y})$ $(\mathbf{R_x}, \mathbf{R_y})$	$(x^2 - y^2, xy), (yz, xz)$

This is the final form of the table of irreducible representations and is known as a **character table**, as each numerical entry in the table is the character of the transformation matrix representing the symmetry operation. The top row shows the point group symbol and the associated symmetry operations. It is important to recognise that these are symmetry *operations* not symmetry *elements*. The point group C_{3v} does not have two C_3 axes, though it does have two operations associated with the one C_3 axis which is present (in addition to $C_3^3 \equiv E$, which is already in the table). On the other hand, as each mirror plane generates only one operation "$3\sigma_v$" means three reflection operations, but this happens to correspond to three mirror planes. The body of the table shows the characters of the transformation matrices and the basis sets on which they are derived. Each of the rows of integers is an irreducible representation for the point group.

Two additional columns have also been added. The left hand column contains **Mulliken symbols**, which are shorthand **symmetry labels** (A_1, A_2, E) for the row of characters with which they are associated. A fuller explanation of how these symmetry labels are arrived at is given in the next section, but at this stage note that "E" is a symmetry label and is different to "*E*" which is the identity symmetry operation! The right hand column lists what are known as "binary combinations" or "binary functions" which are important in the analysis of Raman spectra and *d*-orbitals and whose relevance will be explained in later parts of this book (*Chapters 4* and *9*).

While it is relevant for you to appreciate how different irreducible representations can be obtained using either vectors or atomic orbitals (*Section 2.3*), it is the reverse process which is more important. Thus, subsequent chapters of this book will show how the irreducible representations, through their symmetry labels, can be used to describe either molecular vibrations or atomic / molecular orbitals.

2.5 SYMMETRY LABELS

Mulliken symbols (symmetry labels) provide a shorthand way of describing an irreducible representation. They are arrived at by considering whether or not a representation is symmetric (1) or anti-symmetric (−1) with respect to a series of symmetry operations. A listing of the symbols and their origin are given in Table 2.4.

Table 2.4 Mulliken symbols and their origin.

Symbol	Origin
A or B	singly degenerate (corresponds to a 1 x 1 matrix in character table)
E	doubly degenerate (2 x 2 matrix)
T	triply degenerate (3 x 3 matrix)
A	symmetric with respect to (w.r.t) rotation about the main C_n axis (1 in character table)
B	anti-symmetric w.r.t rotation about the main C_n axis (−1 in character table)
Subscript 1	symmetric w.r.t C_2 perpendicular to C_n, or σ_v if no C_2 present
Subscript 2	anti-symmetric w.r.t C_2 perpendicular to C_n, or σ_v if no C_2 present
g	symmetric w.r.t inversion
u	anti-symmetric w.r.t inversion
'	symmetric w.r.t σ_h
''	anti-symmetric w.r.t σ_h

For example, the A_1 label (*see* Table 2.3) refers to a unique entity (this will manifest itself as a 1×1 matrix, i.e. a character of 1, under the identity operation E), which is symmetric with respect to both rotation about the main axis (A; character 1 in table) and reflection (subscript 1; character 1 in Table). A second example, which will be familiar from other aspects of inorganic chemistry, is the designation of the d_{xy}, d_{xz} and d_{yz} atomic orbitals as "t_{2g}" when placed in an octahedral crystal field (*see* the character table for O_h symmetry in Appendix 5). The use of lower case labels, rather than the capitals which have been shown in Table 2.4, is a convention of atomic / molecular orbital descriptions but otherwise the meanings are the same. The t_{2g} label indicates that these three *d*-orbitals form a set of three that behave collectively as a trio (manifest as a 3×3 transformation matrix and a character of 3 under E in the table), which are anti-symmetric with respect to C_2 (subscript 2) and symmetric with respect to inversion (g) (Fig. 2.3).

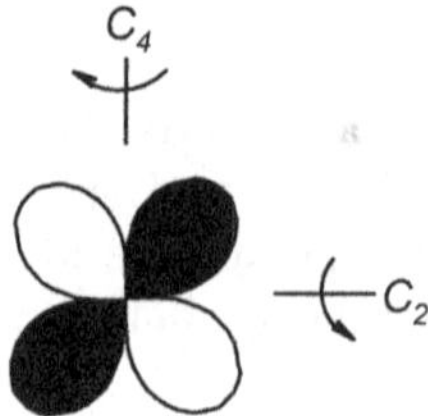

Fig. 2.3 The d_{xy}, d_{xz} and d_{yz} orbitals are anti-symmetric with respect to a C_2 axis at right angles to the principal axis of the O_h point group (C_4) but symmetric with respect to an inversion centre i (**note**: the d-orbitals individually do not have C_4 symmetry).

The labels g and u are only relevant where a centre of inversion is present. Thus, in a tetrahedral crystal field the d_{xy}, d_{xz} and d_{yz} orbitals have label t_2 as there is no inversion centre associated with the T_d point group.

In general, Mulliken symbols can be taken at face value i.e. they are just descriptive labels whose origin is of secondary importance. However, as far as the analysis of vibrational spectra and bonding with which this book is concerned, you will need to know the origin of the following, which are by far the most important of the labels:

- A, B, E, T which relate to **degeneracy**.
- g, u which relate to symmetry with respect to inversion.

2.6 SUMMARY

- a group is a collection of objects e.g. symmetry operations, which obey certain rules.
- a transformation matrix is the mathematical way of describing the effect of performing a symmetry operation.
- a representation of a point group is a collection of transformation matrices which replicate the behaviour of the group's symmetry operations.
- the representation can be simplified by using the character of each of the transformation matrices.
- the character of a matrix is the sum of the leading diagonal elements.
- the simplest sets of integers (characters) that act as representations are called irreducible representations.
- these can (in part) be derived by looking at the effects of symmetry operations on the translational and rotational vectors ($\mathbf{T}_{x,y,z}$, $\mathbf{R}_{x,y,z}$) which are termed basis sets.
- irreducible representations can be described by symmetry labels.
- a character table collects together the symmetry operations of a point group, the irreducible representations of the group, their symmetry labels and the basis sets on which the irreducible representations are based.

PROBLEMS

Answers to all problems marked with * are given in Appendix 4.

1*. Determine the characters of the matrices **A**, **B** and **C**, where **C** = **AB**.

$$\mathbf{A} = \begin{bmatrix} 1 & 2 & 1 \\ 3 & 1 & 1 \\ 1 & 0 & 2 \end{bmatrix} \qquad \mathbf{B} = \begin{bmatrix} 2 & 0 & 2 \\ 1 & -2 & 1 \\ 0 & 3 & 3 \end{bmatrix}$$

2*. What is the 5 × 5 matrix which describes the transformation of the five *d*-orbitals under the operation C_4^1 ? Take the rotation axis to lie along *z*.

3. Consider the five *d*-orbitals as a single basis set. Write out the four 5 × 5 matrices which represent the symmetry operations of the C_{2v} point group applied to this basis set (C_2 lies along *z*).

$$\begin{bmatrix} & & \\ & 5 \times 5 & \\ & & \end{bmatrix} \begin{bmatrix} d_{z^2} \\ d_{x^2-y^2} \\ d_{xy} \\ d_{xz} \\ d_{yz} \end{bmatrix} = \begin{bmatrix} \\ \\ \\ \\ \\ \end{bmatrix}$$

What are the characters of each of the transformation matrices ?

4. Write a matrix for the transformation of the six Cartesian vectors of O_2 (three per oxygen) under the operation of inversion.

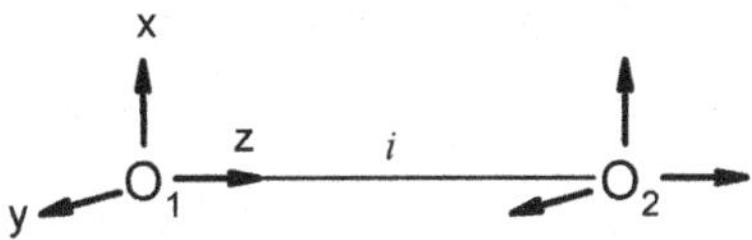

What is the character of this matrix ?

PART II

APPLICATION OF GROUP THEORY TO VIBRATIONAL SPECTROSCOPY

3
Reducible Representations

In this opening chapter of Part II you will see how the techniques of Part I can be applied to the analysis of vibrational spectra. The translations and rotations of a whole molecule were used in Chapter 2 as the basis for determining the irreducible representations of the point group. This is not the total extent to which symmetry governs the movements of the atoms within a molecule, as the vibrational motions – bond stretches and bends – must also obey the point group symmetry.

The problem we will tackle in this chapter is to predict the vibrational spectrum of the simple molecule SO_2 (C_{2v}), using group theory and the demands of point group symmetry to identify which vibrational modes are allowed. This will give you the basic tools required for this type of analysis, though we will return to a more complex example in Chapter 5.

The strategy we will follow is to:

- use three vectors on each atom, to allow the atoms to move independently in each direction *x, y, z*.
- generate a reducible representation of the point group using these nine vectors (three for each of the three atoms in SO_2).
- convert the reducible representation to the sum of a series of irreducible representations.
- identify which of these irreducible representations describe molecular translations and rotations and eliminate these.
- Hence, by difference, identify the symmetry labels associated with the molecular vibrations.

The strategy will unravel as each of these steps is addressed.

3.1 REDUCIBLE REPRESENTATIONS

We will begin by placing three vectors on each atom in the molecule. These vectors allow each atom to move independently along each of the Cartesian axes *x, y, z*. Following the methodology of Chapter 2, we will use these nine vectors (in general, 3 vectors per N atoms = 3N vectors) as a basis set to generate a representation of the point group. From Chapter 2 you will know that this representation originates from the characters of the transformation matrices associated with each of the C_{2v} symmetry operations [E, C_2, $\sigma(xz)$, $\sigma(yz)$] applied to the nine vectors, so our first requirement is to identify these matrices.

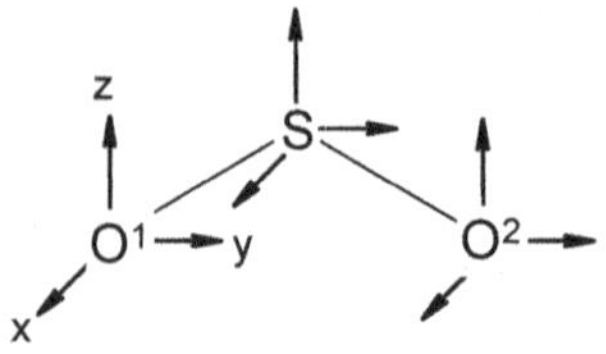

The matrices that are needed are 9 × 9, which shows the potential complexity of the approach. For example, a still relatively simple molecule with 6 atoms would require a series of 18 × 18 matrices. Luckily, shortcuts in this process exist which will be revealed after the four 9 × 9 matrices for SO_2 have been identified.

The matrix which represents the identity operation E is straightforward as all the vectors are unmoved by this operation:

$$\begin{bmatrix} 1&0&0&0&0&0&0&0&0\\ 0&1&0&0&0&0&0&0&0\\ 0&0&1&0&0&0&0&0&0\\ 0&0&0&1&0&0&0&0&0\\ 0&0&0&0&1&0&0&0&0\\ 0&0&0&0&0&1&0&0&0\\ 0&0&0&0&0&0&1&0&0\\ 0&0&0&0&0&0&0&1&0\\ 0&0&0&0&0&0&0&0&1 \end{bmatrix} \begin{bmatrix} S_x\\ S_y\\ S_z\\ O_x^1\\ O_y^1\\ O_z^1\\ O_x^2\\ O_y^2\\ O_z^2 \end{bmatrix} = \begin{bmatrix} S_x\\ S_y\\ S_z\\ O_x^1\\ O_y^1\\ O_z^1\\ O_x^2\\ O_y^2\\ O_z^2 \end{bmatrix}$$

The effect of the C_2 operation on three of the nine vectors is shown diagrammatically below:

C_2:

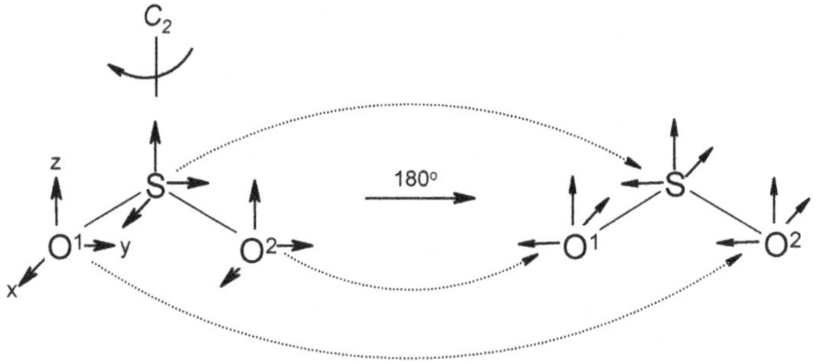

The transformation matrix which describes the movement of all nine vectors under this operation is:

$$\begin{bmatrix} -1 & 0 & 0 & 0 & 0 & 0 & 0 & 0 & 0 \\ 0 & -1 & 0 & 0 & 0 & 0 & 0 & 0 & 0 \\ 0 & 0 & 1 & 0 & 0 & 0 & 0 & 0 & 0 \\ 0 & 0 & 0 & 0 & 0 & 0 & -1 & 0 & 0 \\ 0 & 0 & 0 & 0 & 0 & 0 & 0 & -1 & 0 \\ 0 & 0 & 0 & 0 & 0 & 0 & 0 & 0 & 1 \\ 0 & 0 & 0 & -1 & 0 & 0 & 0 & 0 & 0 \\ 0 & 0 & 0 & 0 & -1 & 0 & 0 & 0 & 0 \\ 0 & 0 & 0 & 0 & 0 & 1 & 0 & 0 & 0 \end{bmatrix} \begin{bmatrix} S_x \\ S_y \\ S_z \\ O_x^1 \\ O_y^1 \\ O_z^1 \\ O_x^2 \\ O_y^2 \\ O_z^2 \end{bmatrix} = \begin{bmatrix} -S_x \\ -S_y \\ S_z \\ -O_x^2 \\ -O_y^2 \\ O_z^2 \\ -O_x^1 \\ -O_y^1 \\ O_z^1 \end{bmatrix}$$

Similarly, for reflection in the *xz* plane, $\sigma(xz)$:

$$\begin{bmatrix} \mathbf{1} & 0 & 0 & 0 & 0 & 0 & 0 & 0 & 0 \\ 0 & \mathbf{-1} & 0 & 0 & 0 & 0 & 0 & 0 & 0 \\ 0 & 0 & \mathbf{1} & 0 & 0 & 0 & 0 & 0 & 0 \\ 0 & 0 & 0 & \mathbf{0} & 0 & 0 & 1 & 0 & 0 \\ 0 & 0 & 0 & 0 & \mathbf{0} & 0 & 0 & -1 & 0 \\ 0 & 0 & 0 & 0 & 0 & \mathbf{0} & 0 & 0 & 1 \\ 0 & 0 & 0 & 1 & 0 & 0 & \mathbf{0} & 0 & 0 \\ 0 & 0 & 0 & 0 & -1 & 0 & 0 & \mathbf{0} & 0 \\ 0 & 0 & 0 & 0 & 0 & 1 & 0 & 0 & \mathbf{0} \end{bmatrix} \begin{bmatrix} S_x \\ S_y \\ S_z \\ O_x^1 \\ O_y^1 \\ O_z^1 \\ O_x^2 \\ O_y^2 \\ O_z^2 \end{bmatrix} = \begin{bmatrix} S_x \\ -S_y \\ S_z \\ O_x^2 \\ -O_y^2 \\ O_z^2 \\ O_x^1 \\ -O_y^1 \\ O_z^1 \end{bmatrix}$$

SAQ 3.1 : What is the transformation matrix associated with the effect of the symmetry operation $\sigma(yz)$ on each of the nine vectors of SO_2?

Answers to all SAQs are given in Appendix 3.

The characters of the four transformation matrices are 9 (E), −1 (C_2), 1 ($\sigma(xz)$; highlighted above) and 3 ($\sigma(yz)$). The representation for the point group C_{2v} is thus:

	E	C_2	$\sigma(xz)$	$\sigma(yz)$
Γ_{3N}	9	−1	1	3

Γ_{3N} (spoken as "gamma 3N") is shorthand for "the representation of the point group based on the 3N vectors as basis set".

It is not, however, necessary to write out the complete transformation matrices to generate Γ_{3N} as we have done above, as it is only the integers which lie on the leading diagonal of the matrix which contribute to its character. Inspection of any of the four transformation matrices reveals the following basic rules:

- if a vector is not moved by an operation it contributes 1 to χ
  e.g. all vectors under the operation *E*.

- if a vector is moved to a new location by an operation it contributes 0 to χ
  e.g. $O_x^1 \rightarrow -O_x^2$ under C_2.

- if a vector is reversed by an operation it contributes −1 to χ
  e.g. $S_y \rightarrow -S_y$ under $\sigma(xz)$.

Thus, the character of the transformation matrix for *E* is simply 9 × 1 (nine vectors all unmoved). For C_2, vectors associated with the two oxygens move to new locations (6 × 0), the *z* vector on S is unmoved (1 × 1) while the S_x and S_y are reversed (2 × −1), giving $\chi = 0 + 1 - 2 = -1$.

SAQ 3.2 : *Use the above method to show that the character of the transformation matrix for* $\sigma(yz)$ (*SAQ 3.1*) is 3.

This approach is entirely valid, but becomes more complex for many other point groups. We will return to this problem and elaborate on the solution in Section 3.4.

While Γ_{3N} is a representation of the point group it is not one of the representations derived in Chapter 2 based on the simple translational or rotational vectors. It is a more complex set of integers and is known as a **reducible representation**. This is because it is, in fact, a sum of the irreducible representations already derived.

The integers 9, −1, 1, 3 as a representation are arrived at by the following sum of the simple irreducible representations:

$$\Gamma_{3N} = 3A_1 + A_2 + 2B_1 + 3B_2$$

This can be confirmed as follows, starting from the character table for C_{2v}:

C_{2v}	*E*	C_2	$\sigma(xz)$	$\sigma(yz)$
A_1	1	1	1	1
A_2	1	1	−1	−1
B_1	1	−1	1	−1
B_2	1	−1	−1	1

By multiplying each irreducible representation by the appropriate coefficient and then summing each column we get:

C_{2v}	*E*	C_2	$\sigma(xz)$	$\sigma(yz)$
$3A_1$	3	3	3	3
A_2	1	1	−1	−1
$2B_1$	2	−2	2	−2
$3B_2$	3	−3	−3	3
Γ_{3N}	9	−1	1	3

Before discussing what this means for the vibrational spectrum of SO_2, it is important to show a systematic method for converting a reducible representation like

Γ_{3N} (9, −1, 1, 3) into the sum of various irreducible representations ($3A_1 + A_2 + 2B_1 + 3B_2$). This method involves what is known as the **reduction formula**.

3.2 THE REDUCTION FORMULA

The reduction formula provides a means of converting a reducible representation into a sum of the irreducible representations associated with a particular point group. The formula appears daunting, but is simple to apply with practice as it merely requires you to multiply and add integers.

$$a_i = \frac{1}{g}\sum (n_R \chi_{(R)} \chi_{(IR)}) \qquad \text{sum over all classes of operation}$$

- a_i : number of times an irreducible representation contributes to the reducible representation
- g : the total number of symmetry operations for the point group
- n_R : the number of operations in a particular class of operation
- $\chi_{(R)}$: the character in the reducible representation corresponding to the class of operation
- $\chi_{(IR)}$: the corresponding character in the irreducible representation

An example will show that this is far less difficult than it looks. In the case of SO_2 we must apply the above formula to see how many times each of the irreducible representations for C_{2v} (A_1, A_2, B_1, B_2) contribute to Γ_{3N}. The total number of operations for the point group (g) is 4 (E, C_2, $\sigma(xz)$, $\sigma(yz)$) and each class of operation consists of only one operation (n_R).

> *SAQ 3.3 : What are the total number of operations for the C_{3v} point group ? How many operations are associated with each class ?*
> *See Appendix 5 for the C_{3v} character table.*

The formula must be applied to each of the irreducible representations in turn, so for A_1:

$$\begin{array}{lllll} & E & C_2 & \sigma(xz) & \sigma(yz) \\ A_1 = 1/4\,[& (1 \times 9 \times 1) + & (1 \times -1 \times 1) + & (1 \times 1 \times 1) + & (1 \times 3 \times 1)] = 12/4 = 3 \end{array}$$

- $\chi_{(IR)}$, the character in the irreducible representation A_1 for the operation E
- $\chi_{(R)}$, the character in the reducible representation for the operation E
- n_R, the number of operations associated with E

For the remaining irreducible representations, the formula generates:

$$A_2 = 1/4\,[(1 \times 9 \times 1) + (1 \times -1 \times 1) + (1 \times 1 \times -1) + (1 \times 3 \times -1)] = 4/4 = 1$$

$$B_1 = 1/4[(1 \times 9 \times 1) + (1 \times -1 \times -1) + (1 \times 1 \times 1) + (1 \times 3 \times -1)] = 8/4 = 2$$

$$B_2 = 1/4[(1 \times 9 \times 1) + (1 \times -1 \times -1) + (1 \times 1 \times -1) + (1 \times 3 \times 1)] = 12/4 = 3$$

Thus:

$$\Gamma_{3N} = 3A_1 + A_2 + 2B_1 + 3B_2$$

Care is important in applying this formula as it is easy to make mistakes. Note that the first two numbers in each group of three (n_R and $\chi_{(R)}$) are always the same for all the irreducible representations and the only number that changes is $\chi_{(IR)}$, which is read from the character table. As a final check, the contribution each irreducible representation makes must be an integer value, so if you get a fractional answer, you have made a mistake !

3.3 THE VIBRATIONAL SPECTRUM OF SO_2

What does it mean when we say that, using the 3N vectors as basis set, the representation of the C_{2v} point group is $3A_1 + A_2 + 2B_1 + 3B_2$? The 3N vectors allow the three atoms of SO_2 to move independently in any direction *x, y, z*. However, the symmetry of the C_{2v} point group imposes restrictions on these movements such that only certain movements of the atoms are allowed, i.e. those which conform to the symmetry operations of the point group. In total, nine concerted atomic movements are allowed, corresponding to each of the three atoms moving in any one of three directions (3N movements in general). These nine concerted motions are described by the nine symmetry labels $3A_1 + A_2 + 2B_1 + 3B_2$.

Of the nine concerted movements, you already know that some of them correspond to simple translations and rotations of the whole molecule. Inspection of the point group character table shows the following:

C_{2v}	*E*	C_2	$\sigma(xz)$	$\sigma(yz)$	
A_1	1	1	1	1	$\mathbf{T_z}$
A_2	1	1	−1	−1	$\mathbf{R_z}$
B_1	1	−1	1	−1	$\mathbf{T_x}$, $\mathbf{R_y}$
B_2	1	−1	−1	1	$\mathbf{T_y}$, $\mathbf{R_x}$

There are three translations (the whole molecule moving along *x*, *y* or *z*) and three rotations (the whole molecule rotating about *x*, *y* or *z*) and these are described by the symmetry labels A_1 ($\mathbf{T_z}$), A_2 ($\mathbf{R_z}$), $2B_1$ ($\mathbf{T_x}$, $\mathbf{R_y}$) and $2B_2$ ($\mathbf{T_y}$, $\mathbf{R_x}$). Counting these labels confirms the six translations + rotations. In summary:

$$\Gamma_{3N} = 3A_1 + A_2 + 2B_1 + 3B \qquad (= 3N, \text{ i.e. } 9 \text{ for } SO_2)$$

$$\Gamma_{\text{translation + rotation}} = A_1 + A_2 + 2B_1 + 2B_2 \qquad (= 6)$$

After removing the symmetry labels which describe the translations and rotations of the molecule from the list which describes all the concerted atomic movements (Γ_{3N}), the labels that remain describe the vibrations of the molecule:

$$\Gamma_{vibration} = 2A_1 + B_2 \qquad (= 3N\text{-}6, \text{ i.e } 3 \text{ for } SO_2)$$

All non-linear molecules exhibit 3N-6 vibrational modes and this should be used as a final check that the reduction formula has been applied correctly. For a linear molecule, rotation about the molecular axis leaves the molecule unmoved, so in this case the vibrational spectrum is described by 3N-5 unique modes.

> *SAQ 3.4 : Using three vectors per atom and the contributions to χ of 1, 0, -1 for vectors which are unmoved, moved and reversed, respectively, derive Γ_{3N} and hence Γ_{vib} for WF_5Cl (C_{4v}).*
>
> *See Appendix 5 for the character table for C_{4v}.*

We have established that the vibrational spectrum of SO_2 is described by the symmetry labels $2A_1$ and B_2 – what does this mean ? Primarily, it tells us that there are three vibrational modes, but what do these look like ? At least some, though not all, of the vibrational modes must involve stretching of the S-O bonds, so we will begin by trying to identify the possibilities here. We will use single-headed arrows along each S-O bond as "stretching vectors", which allow each bond to stretch independently; the symmetry of the point group will determine how the stretches of each bond are coupled together.

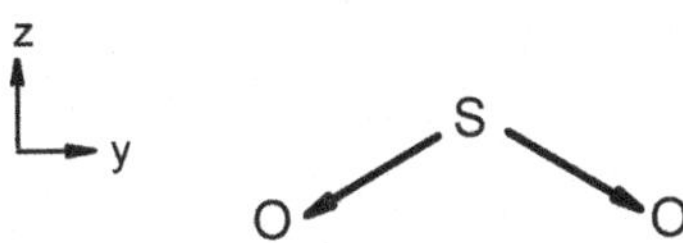

By using these two stretching vectors as a basis set we can determine a representation for the point group:

	E	C_2	$\sigma(xz)$	$\sigma(yz)$
$\Gamma_{S\text{-}O\ stretch}$	2	0	0	2

The integers 2, 0, 0, 2 are the characters of the relevant transformation matrices, which have been generated without writing out the 2×2 matrices by simply counting 1, 0 or −1 depending on whether the vector is unmoved, moved or reversed under each symmetry operation. The representation $\Gamma_{S\text{-}O}$ is a reducible representation which can easily be shown to reduce to $A_1 + B_2$ using the reduction formula. These symmetry labels describe the two ways in which the stretching of each S-O bond can be combined, and correspond to the symmetric and anti-symmetric stretches:

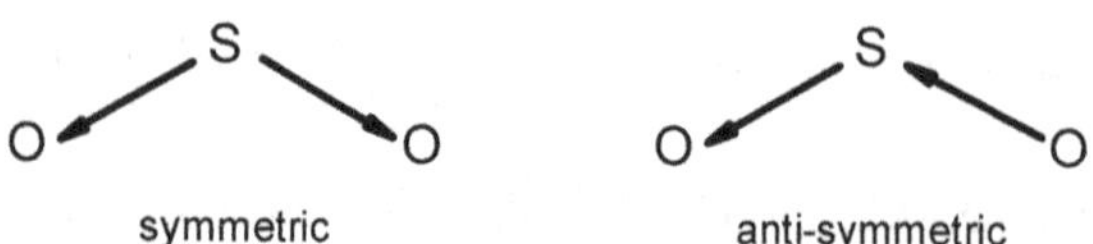

Each pair of vectors describes a unique stretching mode, so now each pair of vectors must be treated as a complete unit. Using each of these two stretching modes in turn as a basis set, we can assign the appropriate symmetry label:

	E	C_2	$\sigma(xz)$	$\sigma(yz)$	
$\Gamma_{\text{S-O symm}}$	1	1	1	1	$= A_1$

Although the two single-headed arrows swap places under the operation C_2, *the pair of vectors as a single unit remains indistinguishable from the original* and hence counts 1; the same applies to the effect of $\sigma(xz)$.

	E	C_2	$\sigma(xz)$	$\sigma(yz)$	
$\Gamma_{\text{S-O antisymm}}$	1	−1	−1	1	$= B_2$

Here, the effect of C_2 (shown below) and $\sigma(xz)$ is to swap the positions of the vectors:

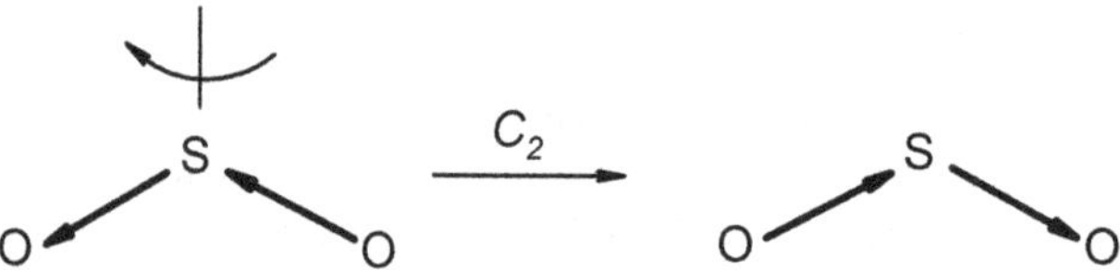

The pair of vectors as a single unit is now the reverse of the original and counts −1.

The remaining vibrational mode does not involve bond stretches – we have accounted for all the possibilities above – but relates to a bending mode in which the O-S-O bond angle is expanded and contracted. This is depicted using a double-headed arrow to represent the bending vector:

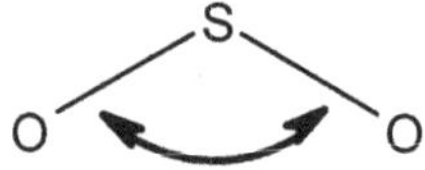

Using this double-headed arrow as the basis, we can derive the representation Γ_{bend}:

	E	C_2	$\sigma(xz)$	$\sigma(yz)$	
Γ_{bend}	1	1	1	1	$= A_1$

As with the effect of C_2 on the pair of stretching vectors describing the symmetric stretch taken as a unit, the effect of C_2 on the double-headed bending arrow swaps the ends of the arrow over *but leaves the double-headed arrow as a complete entity indistinguishable from the original*.

In summary, group theory shows that C_{2v} symmetry allows three vibrational modes for SO_2, two stretching modes of A_1 and B_2 symmetry and a bending mode of A_1 symmetry.

3.4 CHI PER UNSHIFTED ATOM

One of the lessons that should have been learned from *SAQ* 3.4 is that only vectors associated with atoms that do not move under a symmetry operation will have a non-zero contribution to the character of the associated transformation matrix. In other words, when any atom moves its position under a symmetry operation then its three *x*, *y*, *z* vectors also move to new positions and count 0 to the matrix character. This observation provides a simple way to derive Γ_{3N} which does not involve consideration of vectors directly and, moreover, also affords an easy way to deal with point groups which have rotations other than 90° or 180°.

Any atom which does not move under the identity operation *E* has three vectors which also do not move. The transformation matrix which relates to just one atom is 3 × 3 and of the form:

$$\begin{bmatrix} 1 & 0 & 0 \\ 0 & 1 & 0 \\ 0 & 0 & 1 \end{bmatrix} \begin{bmatrix} x \\ y \\ z \end{bmatrix} = \begin{bmatrix} x \\ y \\ z \end{bmatrix} \qquad \chi_{\text{unshifted atom}} = 3$$

The character of this matrix is termed "chi per unshifted atom".

3 × 3 matrices will form part of the overall transformation matrix for the operation *E*, with a 3 × 3 sub-matrix for each atom that does not move. For example, with reference to SO_2:

$$\begin{bmatrix} 1 & 0 & 0 & & & & & & \\ 0 & 1 & 0 & & & & & & \\ 0 & 0 & 1 & & & & & & \\ & & & 1 & 0 & 0 & & & \\ & & & 0 & 1 & 0 & & & \\ & & & 0 & 0 & 1 & & & \\ & & & & & & 1 & 0 & 0 \\ & & & & & & 0 & 1 & 0 \\ & & & & & & 0 & 0 & 1 \end{bmatrix} \begin{bmatrix} S_x \\ S_y \\ S_z \\ O^1_x \\ O^1_y \\ O^1_z \\ O^2_x \\ O^2_y \\ O^2_z \end{bmatrix} = \begin{bmatrix} S_x \\ S_y \\ S_z \\ O^1_x \\ O^1_y \\ O^1_z \\ O^2_x \\ O^2_y \\ O^2_z \end{bmatrix} \qquad \chi = 9$$

All other entries in the matrix, irrespective of their value (they happen to be all 0 in the case above), can be ignored as they do not affect the leading diagonal. We can thus determine the character of the above 9 × 9 transformation matrix i.e. the character in the reducible representation for *E*, by first determining the number of atoms that do not move under the operation *E* (in the case of SO_2, 3) then multiplying this number by the contribution $\chi_{\text{unshifted atom}}$ ($\chi_{\text{u.a.}}$ for short) i.e. 3 × 3 = 9. This is a rather trivial example of the approach as it is obvious that under the identity operation all nine vectors are unmoved, but its value will become more apparent as we look at other symmetry operations.

In the case of inversion, the three vectors on any atom that doesn't move (there can only be one such atom) all invert, so the 3 × 3 matrix which describes this is:

$$\begin{bmatrix} -1 & 0 & 0 \\ 0 & -1 & 0 \\ 0 & 0 & -1 \end{bmatrix} \begin{bmatrix} x \\ y \\ z \end{bmatrix} = \begin{bmatrix} -x \\ -y \\ -z \end{bmatrix} \qquad \chi_{u.a.} = -3$$

For any atom lying on either a σ_h or σ_v mirror plane, and hence does not move, there will be two vectors which also lie in the mirror plane and are therefore unaffected by the operation; the third vector, which is a right-angles to the mirror plane, is reversed. Using $\sigma(xz)$ as an example:

$$\begin{bmatrix} 1 & 0 & 0 \\ 0 & -1 & 0 \\ 0 & 0 & 1 \end{bmatrix} \begin{bmatrix} x \\ y \\ z \end{bmatrix} = \begin{bmatrix} x \\ -y \\ z \end{bmatrix} \qquad \chi_{u.a.} = 1$$

Dihedral planes appear to differ at first sight, but $\chi_{u.a.}$ is still 1:

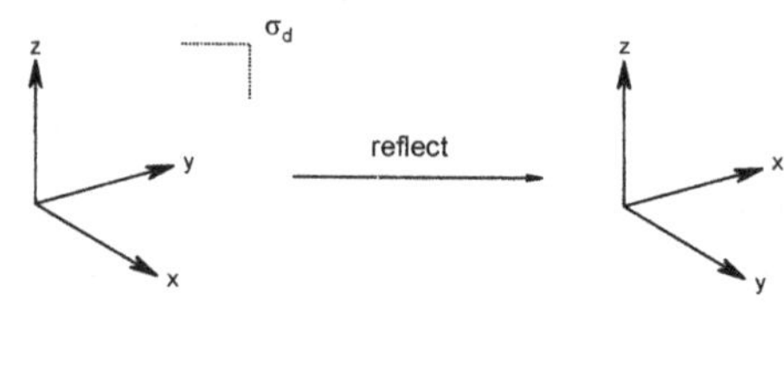

$$\begin{bmatrix} 0 & 1 & 0 \\ 1 & 0 & 0 \\ 0 & 0 & 1 \end{bmatrix} \begin{bmatrix} x \\ y \\ z \end{bmatrix} = \begin{bmatrix} y \\ x \\ z \end{bmatrix} \qquad \chi_{u.a.} = 1$$

The most useful application of this approach relates to the character of a transformation matrix associated with a rotation. In the simple example of SO_2 (C_{2v}), we counted 1 for a vector unmoved by the rotation, −1 if the vector reversed direction and 0 if it moved to a new location. This, however, is a rather special case and in general the analysis is more complex. You have already seen (*Section 2.3*) that under certain rotations, some vectors act together e.g. a pair of vectors are transformed onto a combination of themselves.

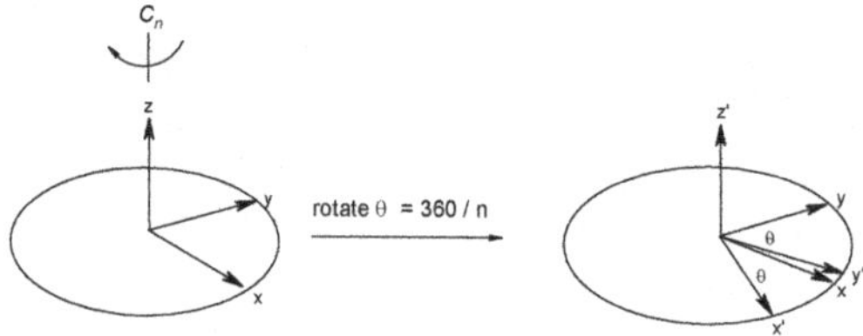

As shown in Section 2.3:

$$x' = \cos\theta\,(x) - \sin\theta\,(y)$$
$$y' = \sin\theta\,(x) + \cos\theta\,(y)$$
$$z' = z$$

The vector along z, which coincides with the rotation axis, is unmoved. The transformation matrix for any atom unmoved by C_n is:

$$\begin{bmatrix} \cos\theta & -\sin\theta & 0 \\ \sin\theta & \cos\theta & 0 \\ 0 & 0 & 1 \end{bmatrix} \begin{bmatrix} x \\ y \\ z \end{bmatrix} = \begin{bmatrix} x' \\ y' \\ z' \end{bmatrix} \qquad \chi_{u.a.} = 1 + 2\cos\theta$$

For a C_2 axis, the contribution to the character of the transformation matrix when a vector rotates through 180° i.e. reverses, is –1 as this is the value of cos180, and is the value we have used earlier in the example of SO_2.

Improper rotations follow a similar pattern for the rotation component, but the additional reflection in σ_h (xy plane) reverses the direction of the vector along z :

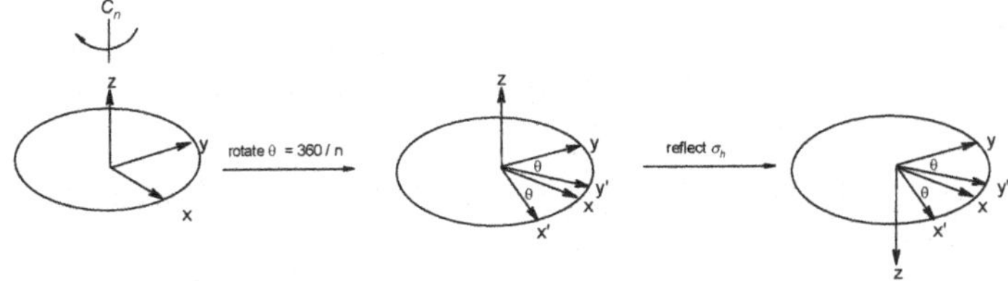

The transformation matrix for any atom which does not move under an S_n rotation is therefore:

$$\begin{bmatrix} \cos\theta & -\sin\theta & 0 \\ \sin\theta & \cos\theta & 0 \\ 0 & 0 & -1 \end{bmatrix} \begin{bmatrix} x \\ y \\ z \end{bmatrix} = \begin{bmatrix} x' \\ y' \\ z' \end{bmatrix} \qquad \chi_{u.a.} = -1 + 2\cos\theta$$

SAQ 3.5 : What is $\chi_{u.a}$ for an S_4 axis ?

In summary:

Table 3.1 Contributions of $\chi_{u.a.}$ for all symmetry operations.

operation	$\chi_{u.a.}$
E	3
i	–3
σ	1
C_n	1 + 2cos(360 / n)
S_n	–1 + 2cos(360 / n)

The example of SO_2 will show how this approach is applied:

C_{2v}	E	C_2	$\sigma(xz)$	$\sigma(yz)$
unshifted atoms	3	1	1	3
$\times \chi_{\text{unshifted atom}}$	3	−1	1	1
Γ_{3N}	9	−1	1	3

As it is easier to visualise the movement of N atoms, rather than 3N vectors, under a given symmetry operation, the use of $\chi_{u.a.}$ provides a much simpler route to generating Γ_{3N}. Note that this methodology can only be used to generate Γ_{3N}. It is *not* applicable to representations using stretching and bending vectors as basis sets.

SAQ 3.6 : Use the approach of "chi per unshifted atom" to generate the representation Γ_{3N} for WF_5Cl (C_{4v}) (see SAQ 3.4).

3.5 SUMMARY

- the 3N vectors associated with N atoms in a molecule can be used to generate a representation for the point group, Γ_{3N}.
- it is easier to generate Γ_{3N} by identifying the atoms that are unmoved under a given symmetry operation and multiplying this number by $\chi_{\text{unshifted atom}}$.
- Γ_{3N} is a reducible representation and can be shown to be made up of a combination of the irreducible representations given in the point group character table.
- the reducible representation is converted to the sum of these irreducible representations using the reduction formula:

$$a_i = \frac{1}{g}\sum(n_R \chi_{(R)} \chi_{(IR)})$$

- the symmetry labels for the vibrational modes of a molecule are derived by subtracting those labels that describe the translations and rotations (read from the character table) from Γ_{3N} i.e $\Gamma_{vib} = \Gamma_{3N} - \Gamma_{trans+rot}$.
- irreducible representations (symmetry labels) for the bond stretches are generated by analysing the effect of the point group symmetry operations on single-headed arrows lying along the bonds.
- similarly, double-headed arrows can be used to derive the number and symmetry labels of the bending modes.

PROBLEMS

Answers to all problems marked with * are given in Appendix 4.

1. Reduce the following representations to the sum of a series of irreducible representations for the point group concerned:

*

C_{2v}	E	C_2	$\sigma(xz)$	$\sigma(yz)$
Γ	6	4	–2	0

*

T_d	E	$8C_3$	$3C_2$	$6S_4$	$6\sigma_d$
Γ	9	3	1	3	3

*

C_{3v}	E	$2C_3$	$3\sigma_v$
Γ	15	0	3

D_{3h}	E	$2C_3$	$3C_2$	σ_h	$2S_3$	$3\sigma_v$
Γ	12	0	–2	4	–2	2

2. Derive Γ_{3N} and hence Γ_{vib} for each of the following:

   * NH_3 * *cis*-N_2H_2 *trans*-N_2H_2 * SO_3 MoF_5

   (For cis-N_2H_2, take the molecular plane to be yz).

3*. Derive Γ_{3N} and hence Γ_{vib} for the planar fumarate dianion (C_{2h}):

4

Techniques of Vibrational Spectroscopy

To analyse the vibrational spectrum of a molecule fully usually requires the application of two techniques – infrared and Raman spectroscopies. This chapter will briefly review the background to these techniques, with particular emphasis on how they relate to the analysis of atomic motion using group theory. It will not provide a *detailed* treatment of either the background or the experimental methodology surrounding these two forms of spectroscopy, as these are available though specialist texts.*

4.1 GENERAL CONSIDERATIONS

The vibration of a bond connecting two atoms can be modelled by considering two spheres of mass m_1 and m_2 joined by a spring whose strength is measured by a *force constant*, k; the stiffer the spring (i.e. the stronger the bond between the atoms) the larger the value of k. In measuring a vibrational spectrum, energy is absorbed by the bond and its vibrational motion increases or, more accurately, the molecule moves from one vibrational energy level to another; usually, this is from the ground vibrational state to the first excited state. The energy that is required to bring about this transition is given by:

$$\mathrm{E} = h\nu \qquad \text{where } \nu = \frac{1}{2\pi}\sqrt{\frac{k}{\mu}} \qquad \text{(eqn. 4.1)}$$

μ is referred to as the *reduced mass* and is given by $m_1m_2/(m_1 + m_2)$.

The energy of radiation required to move molecules between vibrational energy levels is *ca.* 10^{-20} J, which corresponds to a vibrational frequency (the frequency with which the spring oscillates) of *ca.* 10^{13} s^{-1} (10^{13} Hz).

---

* For example, C N Banwell and E M McCash, *Fundamentals of Molecular Spectroscopy*, McGraw Hill, 4th Edition, 1994.

i.e. $\nu = E / h$

$$= 10^{-20}\,\text{J} / 6.63 \times 10^{-34}\,\text{Js} \quad \approx 10^{13}\,\text{s}^{-1}.$$

This energy lies in the infrared part of the electromagnetic spectrum. For vibrational spectra, energy is usually quoted in terms of wavenumber ($1/\lambda$) and has units of cm^{-1}.

$$c = \nu\lambda \qquad \text{from which :} \quad 1/\lambda = \nu/c$$

The vibrational spectrum of a molecule is usually recorded between 4000 and 200 cm^{-1}. For example, a frequency of 10^{13} s^{-1} is equivalent to $10^{13} / 3 \times 10^{10} = 333$ cm^{-1}.

The following key points should be noted from equation 4.1:

- strong bonds (large k) absorb energy at higher wavenumber than weak bonds (small k).
- vibrations involving atoms of high mass (one or both of m_1 or m_2 large) are observed at lower wavenumber than those involving light atoms.

Both of these general observations are exemplified in Table 4.1. As the strength of the C-C bond decreases from a bond order of three to a bond order of one, the vibrational energy moves from > 2000 cm^{-1} to *ca.* 1200 cm^{-1}. An "inorganic" example of the same phenomenon is the lowered stretching energy of carbon monoxide attached to a metal. Terminal CO groups have the strongest C-O bond and require the highest stretching energy. Bridging CO groups, in which the C-O bond is weakened by back-donation of electron density from more than one metal to a π^* orbital on CO, give rise to vibrational bands at lower energies; C=O in simple aldehydes and ketones comes at a still lower energy.

Table 4.1 Representative stretching energies (cm^{-1}) for differing bond types.

Bond	Typical energy	Bond	Typical energy
C≡C	2220	C-F	1050
C=C	1650	C-Cl	725
C-C	1200	C-Br	650
		C-I	550
terminal CO	2000		
bridging CO	1800		
C=O	1700		

The spectra of organic molecules i.e. compounds involving only the light atoms C, H, O, N etc are usually recorded between 4000 and 600 cm^{-1}, while stretches involving heavier elements e.g. metals, come in the *ca.* 400 - 200 cm^{-1} region. Spectra are rarely recorded below 200 cm^{-1}, because, apart from the cost of instrumentation, the only vibrational modes which generally occur in this region are lattice modes i.e. vibrations of neighbouring molecules within a solid lattice.

One final general observation concerns stretching and bending modes:

- stretches require a higher energy than bends.

A bond stretch requires both compression and elongation of a bond, which is controlled by the electron density lying along it. In contrast, a bending mode distorts a bond angle i.e. the space between bonds, where electron density is relatively low.

4.2 INFRARED SPECTROSCOPY

The most common way of recording a vibrational spectrum is by directly exciting the vibrational modes, using a source of radiation which exactly matches the energy gaps between vibrational energy states. This radiation lies in the infrared part of the electromagnetic spectrum and hence the technique is called infrared spectroscopy. In a conventional dispersive infrared experiment, the sample is scanned stepwise by infrared radiation across a range of frequencies and those transitions which give rise to excitation of vibrational modes are recorded sequentially in the spectrum. Currently, this type of instrumentation has been largely replaced by spectrometers that irradiate the sample with a broad range of infrared frequencies simultaneously. The resultant mixture of transmitted radiation is subsequently deconvoluted using Fourier transform techniques to identify which frequencies have been absorbed.

The absorption of infrared energy is, however, subject to a dipole moment change when the molecule is excited between the two vibrational energy levels. If there is no such change, no absorption of radiation occurs and a vibrational mode is said to be *infrared inactive*. The following selection rule applies:

- a vibrational mode is only infrared active if it results in a change in dipole moment on excitation.

For example, the symmetric stretch of linear CO_2 ($D_{\infty h}$) does not involve a change in dipole as the two oxygen atoms are symmetrically disposed about the central carbon atom in both ground and excited states; the molecule retains an inversion centre throughout the excitation process. On the other hand, the anti-symmetric stretch causes the molecule to lose its inversion centre on excitation and there is clearly a change in dipole as one oxygen moves away from carbon while the other gets closer (Fig. 4.1). The symmetric stretch is thus infrared inactive and will not be seen in an infrared spectrum while the anti-symmetric stretch is infrared active and will appear.

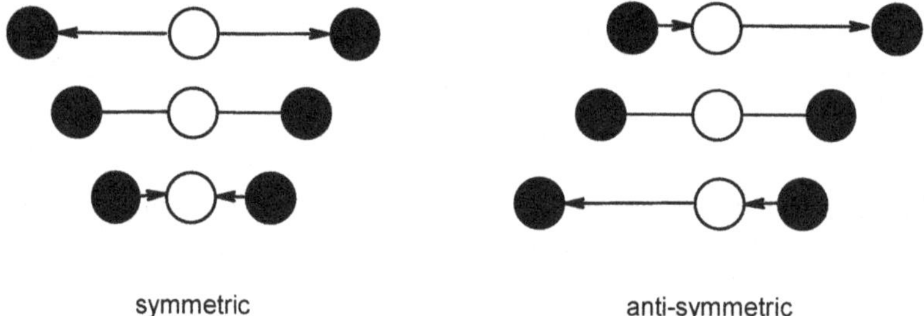

Fig. 4.1 Symmetric and anti-symmetric stretching modes of CO_2.

In Fig. 4.1 it is easy to see for CO_2 when a change in dipole moment occurs and when it does not, though not all cases are so straightforward. Rather than present a detailed theoretical argument at this stage, we will simply state how group theory tells us when a vibrational mode is active:

- a vibrational mode is infrared active if it has the same symmetry as one of the translational vectors ($\mathbf{T_x}$, $\mathbf{T_y}$ or $\mathbf{T_z}$), read from the character table.

A fuller explanation for this rule is given in Chapter 11 (*Section 11.5*).

SAQ 4.1 : The vibrational spectrum of $[PtCl_4]^{2-}$ (square planar, D_{4h}) is given by:

$$\Gamma_{vib} = A_{1g} + B_{1g} + B_{2g} + A_{2u} + B_{2u} + 2E_u$$

Which of these modes are infrared active ? See Appendix 5 for the character table for D_{4h}.

Answers to all SAQs are given in Appendix 3.

4.3 RAMAN SPECTROSCOPY

Unlike infrared spectroscopy, Raman spectroscopy is an indirect technique. It involves irradiating the sample with high intensity radiation from a laser source and measuring the interaction of the sample with this radiation, at right angles to the input direction. Normally, photons of radiation will only excite a molecule if they match the energy spacing between a pair of energy levels, as in infrared spectroscopy. If such a match does not occur, the photon and the sample interact *elastically*, and no energy is transferred. However, in a very small number of cases (*ca.* 1 in 10^7) the interaction is *inelastic*, and energy is transferred between the photon and the sample. To measure these rare events, the intensity of the incident radiation must be high and this criterion is fulfilled by the use of a laser. Moreover, it is difficult to measure these events against a background of the overwhelming majority of elastic collisions, so the radiation scattered at right angles to the direction of incident radiation is measured i.e. against a blank background.

The amount of energy transferred ($\Delta E = h\nu'$) is only a fraction of the total energy of the photon ($h\nu_1$) and corresponds to excitation of a vibrational mode. The energy of the scattered photon is lowered and occurs at new frequency, ν_2. Another possible, though less common event, is the interaction of the photon with a molecule already in an excited vibrational state. Here, energy is lost by the molecule in returning to its vibrational ground state and the frequency (energy) of the scattered photon increases to $\nu_1 + \nu_2$.

A Raman spectrum measures the *change in frequency* of radiation, ($\nu_1 - \nu_2$) or ($\nu_1 + \nu_2$), which corresponds to the frequency (energy) gap between vibrational states, ν'. Unlike infrared spectroscopy, the incident radiation is of a single frequency (monochromatic radiation) and the vibrational spectrum is measured indirectly.

The selection rule for Raman spectroscopy is that:

- a vibrational mode is only Raman active if it results in a change in molecular polarisability.

Polarisability describes the ease or difficulty with which an electron cloud can be distorted. In the case of the symmetric stretch of CO_2, the polarisability changes as both the bonds simultaneously either shorten or lengthen (Fig. 4.1), so while this mode is not infrared active it is Raman active.

In group theory terms, the criterion for change in polarisability is :

- a vibrational mode is Raman active if it has the same symmetry as one of the binary terms (*xy*, *xz*, x^2-y^2 etc), read from the character table.

> *SAQ 4.2 : The vibrational spectrum of* $[PtCl_4]^{2-}$ *(square planar,* D_{4h}*) is given by:*
>
> $$\Gamma_{vib} = A_{1g} + B_{1g} + B_{2g} + A_{2u} + B_{2u} + 2E_u$$
>
> *Which of these modes are Raman active ?*

4.4 RULE OF MUTUAL EXCLUSION

In the case of SO_2 (C_{2v}), whose vibrational spectrum was shown in Chapter 3 to consist of $2A_1$ and B_2 modes, all three bands are observable in both the infrared and Raman experiments (A_1 : symmetry of $\mathbf{T_z}$, x^2, y^2, z^2; B_2 : $\mathbf{T_y}$, *yz*). The complete vibrational spectrum thus contains **coincidences**, bands which occur in both infrared and Raman spectra at essentially the same energy.

Table 4.2 The vibrational spectrum of SO_2.

Infrared (vapour; cm^{-1})	Raman (liquid; cm^{-1})
1362	1336 (depol)
1151	1145 (pol)
518	524 (pol)

The variations which can occur in these data result from both the small error which is inherent in any experiment (*ca.* ± 2 cm^{-1}) and, as in Table 4.2, potentially larger differences when the sample is recorded in two different states for the two techniques e.g. vapour *vs* liquid. In general, if the phase of the sample is the same for both techniques, then the differences in coincident data will be only a few wavenumbers in magnitude.

The infrared and Raman activities of the vibrational modes of $[PtCl_4]^{2-}$ (*SAQ 4.1*, *SAQ 4.2*) differ from those of SO_2 and illustrate an important point. Here, a mode is either infrared or Raman active but not both; there are no coincidences between the two spectra. A_{2u} ($\mathbf{T_z}$) and $2E_u$ ($\mathbf{T_x}$, $\mathbf{T_y}$) are infrared active, while A_{1g} ($x^2 + y^2$, z^2), B_{1g} ($x^2 - y^2$) and B_{2g} (*xy*) are all Raman active. Note the very low energies of vibration, due to the high masses of both platinum and chlorine (Table 4.3).

Table 4.3 The vibrational spectrum of $[PtCl_4]^{2-}$.

Infrared (solution; cm^{-1})	Raman (solution; cm^{-1})
	332 (pol)
320	
	314 (depol)
183	
	170 (depol)
93	

This is an example of a general phenomenon known as the **rule of mutual exclusion**:

- any molecule containing a centre of inversion as a symmetry element will not show any coincidences between its infrared and Raman spectra.

SAQ 4.3 : *The vibrational spectrum of difluorodiazine, FN=NF, is given below. Which geometric isomer is it ?*

Infrared (gas, cm^{-1})	Raman (gas, cm^{-1})
	1636 (pol)
	1010 (pol)
989	
	592 (pol)
412	
360	

Finally, some word of explanation is required for the terms "pol" and "depol" under the Raman data in Tables 4.2, 4.3 and *SAQ 4.3*. Electromagnetic waves can have any orientation in space. If, in a Raman experiment, the incident radiation is confined to only one plane ("plane-polarised") by the use of a slit, rather than randomly oriented, then the interactions with the molecule under study will vary from one molecular vibration to another. If we consider the symmetrical stretch of a spherically symmetrical molecule such as CCl_4, it can be seen that the polarisability of the molecule will change as all the bonds stretch or contract simultaneously. However, as this expansion or contraction is perfectly symmetrical, the polarisability, though different, remains the same in all directions. It does not matter what orientation in space the CCl_4 molecule adopts, the polarisability of the molecule, as seen by the incident plane-polarised radiation, will always remain the same. The net effect is that, in an interaction with a species undergoing a perfectly symmetrical vibration, plane-polarised radiation remains polarised after scattering by the sample (Fig. 4.2a).

On the other hand, an unsymmetrical stretching mode, e.g. one in which one bond elongates while the others contract, will lead to both a change in polarisability and also a loss in its spherical symmetry. The plane polarised light will see a different degree of polarisability depending how the CCl_4 molecule is oriented toward the incident radiation. The result of this is that, despite starting off as plane-polarised, the wave becomes depolarised after scattering by an unsymmetrical vibrational mode (Fig. 4.2b).

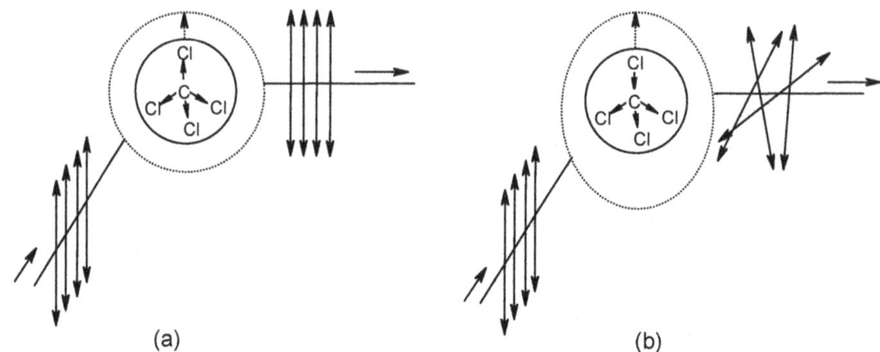

Fig. 4.2 (a) Plane-polarised radiation remains plane-polarised after an interaction with a perfectly symmetrical vibrational mode, and (b) becomes depolarised after interaction with an unsymmetrical mode.

This leads to the following general observations:

- a Raman mode will only be polarised if it is perfectly symmetrical.
- a Raman polarised mode corresponds to a 1 in the character table for all symmetry operations – the totally symmetric representation – and always corresponds to the first of the irreducible representations listed.

As an example, consider the vibrational spectrum of SO_2 (Table 4.2). All three modes, $2A_1$ and B_2, are active in the infrared and Raman, but how can they be assigned? Two of the modes give rise to polarised bands and so must correspond to the A_1 vibrations i.e. the symmetric stretch and the bending mode (*Section 3.3*). The remaining B_2 mode gives rise to the Raman depolarised band. The two A_1 modes can be distinguished by remembering that stretching modes generally come at higher energy than bending modes, so that the complete spectral analysis is:

Infrared (vapour; cm^{-1})	Raman (liquid; cm^{-1})	Assignment
1362	1336 (depol)	B_2 anti-symmetric stretch
1151	1145 (pol)	A_1 symmetric stretch
518	524 (pol)	A_1 bend

4.5 SUMMARY

- the complete vibrational spectrum of a species usually requires measurement of both its infrared and Raman spectra.
- an infrared spectrum is recorded by irradiating a sample with infrared radiation, which causes a transition between one vibrational energy level and another.
- a vibrational mode will only be infrared active if excitation between vibrational energy levels leads to a change in dipole moment.
- a mode is infrared active if it has the same symmetry as any one of the translational vectors, $\mathbf{T_x}$, $\mathbf{T_y}$ or $\mathbf{T_z}$.
- a Raman spectrum is recorded by noting the change in frequency that occurs when laser radiation interacts inelastically with the molecules of the sample.
- a vibrational mode will only be Raman active if excitation between vibrational energy levels leads to a change in polarisability.
- a mode is Raman active if it has the same symmetry as any of the binary combinations xy, xz, x^2-y^2 etc.
- Raman polarised bands only occur for symmetrical modes, which correspond to the first irreducible representation listed in the character table.
- the rule of mutual exclusion states that any species which has an inversion centre as a symmetry element will show no coincidences between its infrared and Raman spectra.

PROBLEMS

Answers to all problems marked with * are given in Appendix 4.

See Appendix 5 for all the relevant character tables.

1*. The vibrational spectrum of $[ClO_4]^-$ is given by:

$$\Gamma_{vib} = A_1 + E + 2T_2$$

Assign infrared and Raman activities, and Raman polarisation data to each of these modes.

2*. The vibrational spectra (cm^{-1}) of the ions $[BrF_2]^-$ and $[BrF_2]^+$ are given below. Assign spectra **A** and **B** to the appropriate ion, giving your reason.

A	**B**
596 (IR)	715 (IR, R pol)
442 (R pol)	706 (IR, R depol)
198 (IR)	366 (IR, R pol)

3. The vibrational spectrum of N_2 contains a Raman-only band at 2331 cm^{-1}. The ion $[Ru(NH_3)_5(N_2)]^{2+}$ has an N_2 stretching frequency of 2105 cm^{-1} active in both the infrared and Raman spectra, while $[Ru_2(NH_3)_{10}(N_2)]^{4+}$ has a Raman-only active band at 2100 cm^{-1}. Comment on these data.

4. Deduce Γ_{vib} and hence predict the infrared and Raman activities (including polarisation data) for $[FeCl_6]^{3-}$.

5. Using single-headed arrows along each of the C-O bonds, predict the number and IR/R activity of the C-O stretching modes of both *mer*- and *fac*-isomers of $L_3M(CO)_3$ (L = ligand).

The compound $Mo(CO)_3(PPh_2Me)_3$ has bands at 1959, 1853, 1845cm^{-1} in its infrared spectrum. Which isomer is it?

6. Seven-coordinate species can adopt either capped trigonal prism (C_{2v}) or pentagonal bipyramidal (D_{5h}) geometries:

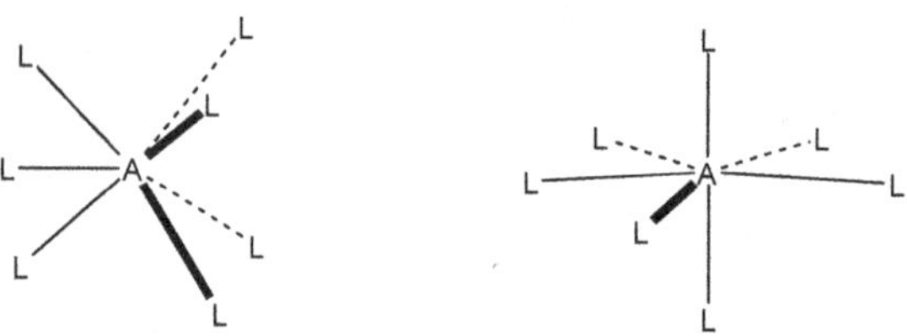

The species $K_4[Mo(CN)_7]$ has been studied by infrared spectroscopy in both the solid state and aqueous solution. Data (cm^{-1}) for the region of the spectrum associated with the C≡N stretching modes are:

Solid : 2119, 2115, 2090, 2080, 2074, 2059.
Solution : 2080, 2040.

Determine Γ_{vib} using the seven C→N stretching vectors as basis set and thus the infrared activities of the ν(C≡N) modes for each of the two geometries. Hence rationalise these data. (Assign the *z* direction to that of the principal axis).

7. $Fe(CO)_4Cl_2$ has infrared bands at 2167, 2126 and 2082 cm^{-1} associated with the carbonyl group stretches. How would you interpret this data ?

5

The Vibrational Spectrum of Xe(O)F_4

This chapter brings together the ideas of Chapters 3 and 4 and summarises the methodology for completely analysing and assigning a vibrational spectrum. We will begin with a more detailed look at the way symmetry labels for stretching and bending modes are determined, followed by a worked example for the vibrational spectrum of Xe(O)F_4. Finally, some comments on the limitations of using group theory to analyse vibrational spectra will be made and the alternative approach, using group frequencies, discussed.

5.1 STRETCHING AND BENDING MODES

In Chapter 3 you saw how to use single- and double-headed arrows to identify the stretching and bending modes for a simple molecule like SO_2. We will now apply the same approach to a more complex example, namely the analysis of the vibrational modes of $[PtCl_4]^{2-}$ (D_{4h}; Fig. 5.1 shows the location of the key symmetry elements):

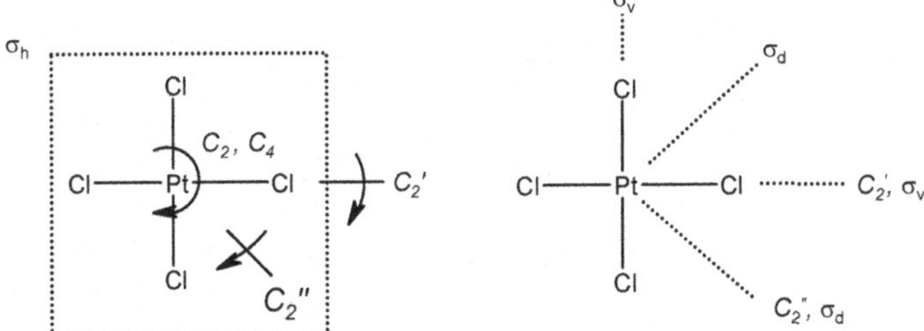

Fig. 5.1 The location of the symmetry elements present in $[PtCl_4]^{2-}$ (D_{4h}).

SAQ 5.1 : Derive Γ_{3N} for $[PtCl_4]^{2-}$ and hence show that:

$$\Gamma_{vib} = A_{1g} + B_{1g} + B_{2g} + A_{2u} + B_{2u} + 2E_u$$

See Appendix 5 for the character table for D_{4h}.

Answers to all SAQs are given in Appendix 3.

The vibrational spectrum for this ion is given below, in which the infrared-active modes are A_{2u}, E_u (*SAQ 4.1*), while the Raman-active modes are A_{1g} (pol), B_{1g} (depol) and B_{2g} (depol) (*SAQ 4.2*):

Infrared (solution; cm^{-1})	Raman (solution; cm^{-1})
	332 (pol)
320	
	314 (depol)
183	
	170 (depol)
93	

As with SO_2, we can identify the symmetry labels for the bond stretches by using single-headed arrows along each Pt-Cl bond. Using these four vectors as basis set, the representation (characters of the transformation matrices) is obtained by counting 1 if the vector is unmoved or 0 if it moves to a new location:

D_{4h}	E	$2C_4$	C_2	$2C_2'$	$2C_2''$	i	$2S_4$	σ_h	$2\sigma_v$	$2\sigma_d$
$\Gamma_{Pt\text{-}Cl}$	4	0	0	2	0	0	0	4	2	0

Note that under the operation C_2, the four vectors appear to reverse on themselves, as shown below for one of the vectors:

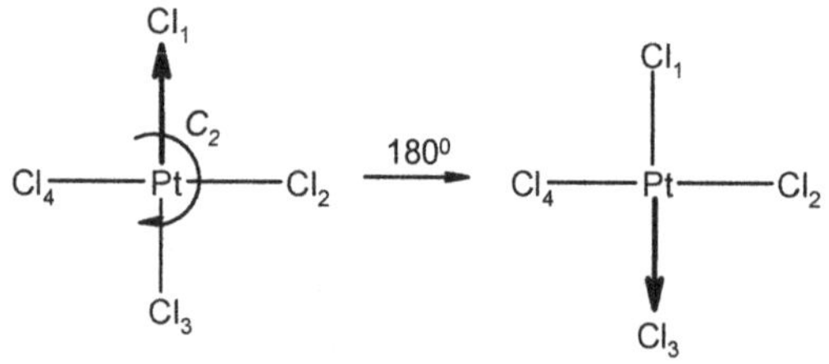

Unlike Γ_{3N}, where vectors along x, y, z were used as basis set, we do not count –1 when a stretching vector *appears to reverse on itself.* In the case of Γ_{3N}, it is clearly possible for an atom to move in both + and – directions and thus any vector can reverse on itself as the movement of the atom reverses. However, in the case of a stretching vector, the locations of the atoms are fixed so the concept of a bond stretching in the opposite direction to its original position has no meaning. The transformation matrix for this C_2 rotation makes it clear why this situation counts 0 towards the matrix character:

$$\begin{bmatrix} 0 & 0 & 1 & 0 \\ 0 & 0 & 0 & 1 \\ 1 & 0 & 0 & 0 \\ 0 & 1 & 0 & 0 \end{bmatrix} \begin{bmatrix} Pt-Cl_1 \\ Pt-Cl_2 \\ Pt-Cl_3 \\ Pt-Cl_4 \end{bmatrix} = \begin{bmatrix} Pt-Cl_3 \\ Pt-Cl_4 \\ Pt-Cl_1 \\ Pt-Cl_2 \end{bmatrix}$$

Using the reduction formula, we can separate $\Gamma_{Pt\text{-}Cl}$ into the following sum:

$$\Gamma_{Pt\text{-}Cl} = A_{1g} + B_{1g} + E_u$$

> *SAQ 5.2 : Use single-headed arrows to determine the symmetries of the N-O stretching modes of $[NO_3]^-$.*
> *Identify the infrared and Raman activities of each mode.*

Our next task is to identify the bending (also referred to as "deformation") modes. There are two possibilities here, depending on if the deformation is in the plane of the atoms ("in-plane modes") or perpendicular to this plane ("out-of-plane modes"). The former are described by double-headed arrows (Fig. 5.2a), as in the case of SO_2 (Section 3.3).

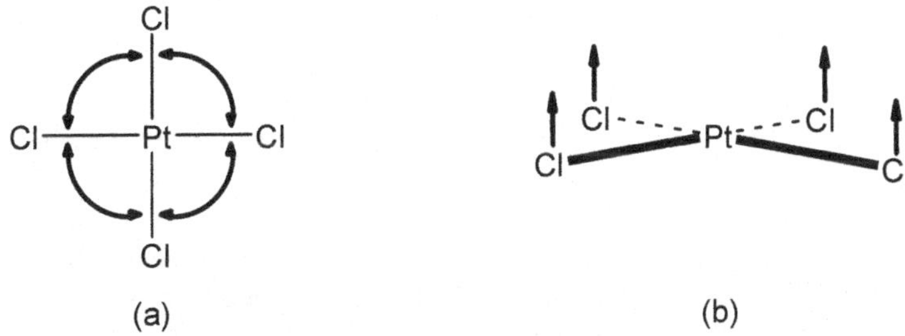

Fig. 5.2 Double- and single-headed arrows describing (a) the in-plane and (b) out-of-plane deformation modes, respectively.

Single-headed arrows are used for the out-of-plane modes (Fig. 5.2b), but note in this case that atoms can move in both + and − directions so reversal of the vector (contributes −1 to the character of the transformation matrix) is possible, along with no movement (1) and movement to a new location (0).

Using these rules, the representation generated by the in-plane bends is:

D_{4h}	E	$2C_4$	C_2	$2C_2'$	$2C_2''$	i	$2S_4$	σ_h	$2\sigma_v$	$2\sigma_d$
$\Gamma_{in\text{-}plane}$	4	0	0	0	2	0	0	4	0	2

Which reduces to:

$$\Gamma_{in\text{-}plane} = A_{1g} + B_{2g} + E_u$$

There are two points to note here. Firstly, under both C_2'' and σ_d, two of the double-headed arrows move to new positions while the other two (those bisected by the axis or plane of symmetry) swap the ends of the double-headed arrow over *but the double-headed arrow as an entity is unmoved.* In this respect, the situation is no different from that encountered in Section 3.3 for the symmetric stretching mode of SO_2. Secondly, we have generated a second mode of A_{1g} symmetry, which cannot be correct as Γ_{vib} has only one A_{1g} term. In fact, this additional A_{1g} mode is an artefact of the group theory approach: the symmetry of the point group would allow all four angles to deform in an identical manner (this perfectly symmetrical mode is what the A_{1g} label describes). However, it is physically impossible for all four bond angles to expand (or contract) at the same time as the four angles must sum to 360°. This second A_{1g} mode is what is known as a **redundancy** and can be discarded.

> *SAQ 5.3 : Use double-headed arrows to determine the symmetries of the N-O in-plane bending modes of $[NO_3]^-$.*
> *Comment on your answer given that $\Gamma_{vib} = A_1' + 2E' + A_2''$.*

We can make a good guess as to what the symmetry labels for the out-of-plane bending modes of $[PtCl_4]^{2-}$ should be, by eliminating the labels describing the stretches and in-plane bends from the total:

$$\Gamma_{vib} = A_{1g} + B_{1g} + B_{2g} + A_{2u} + B_{2u} + 2E_u$$
$$\Gamma_{Pt\text{-}Cl} = A_{1g} + B_{1g} + E_u$$
$$\Gamma_{in\text{-}plane} = B_{2g} + E_u$$

$$\Gamma_{out\text{-}of\text{-}plane} = A_{2u} + B_{2u}$$

To show that this is correct, we can deduce the following representation for the out-of-plane deformations using the set of four single-headed arrows of Fig. 5.2(b) as basis set :

D_{4h}	E	$2C_4$	C_2	$2C_2'$	$2C_2''$	i	$2S_4$	σ_h	$2\sigma_v$	$2\sigma_d$
$\Gamma_{out\text{-}of\text{-}plane}$	4	0	0	–2	0	0	0	–4	2	0

Note that under C_2' two of the single-headed arrows reverse direction (those at right-angles to the axis, count –1) corresponding to a downward movement of the chlorine; the other two single-headed arrows move to new positions (count 0). For σ_h, all four arrows reverse direction (4 × –1).

The representation reduces to :

$$\Gamma_{out\text{-}of\text{-}plane} = E_g + A_{2u} + B_{2u}$$

This does not agree with what we predicted, as it includes an E_g term which is not contained in Γ_{vib}. This is another redundancy, though its origin is less obvious to spot. In fact, E_g is the symmetry label associated with rotations about the x and y axes ($\mathbf{R_x}$, $\mathbf{R_y}$) and, although not part of Γ_{vib}, it does occur in both Γ_{3N} and $\Gamma_{trans+rot}$ (*SAQ 5.1*). The choice of single-headed arrows as a basis for the out-of-plane deformations (Fig. 5.2) happens to also generate two combinations which are rotations of the whole molecule rather than deformations and can thus be discarded from our analysis of $\Gamma_{\text{out-of-plane}}$.

> *SAQ 5.4 : Use single-headed arrows to determine the symmetries of the N-O out-of-plane bending modes of $[NO_3]^-$.*
> *Comment on your answer given the answers to SAQ 5.2 and 5.3*

We are now in a position to fully assign the vibrational spectrum of $[PtCl_4]^{2-}$ (Table 5.1) :

$$\Gamma_{vib} = A_{1g} + B_{1g} + B_{2g} + A_{2u} + B_{2u} + 2E_u \quad (\mathit{SAQ\ 5.1})$$

We begin by noting that A_{1g}, B_{1g} and B_{2g} are Raman-only active, A_{2u} and E_u are infrared-only active and B_{2u} is totally inactive. For the Raman bands, A_{1g} is polarised while the B_{1g}, B_{2g} bands are both depolarised so need to be distinguished. B_{1g} is a Pt-Cl stretch so should come at higher energy than B_{2g}, which is a bending mode.

Table 5.1 The vibrational spectrum of $[PtCl_4]^{2-}$ with assignments.

		Infrared (solution; cm^{-1})	Raman (solution; cm^{-1})		
			332 (pol)	A_{1g}	Pt-Cl stretch
Pt-Cl stretch	E_u	320			
			314 (depol)	B_{1g}	Pt-Cl stretch
in-plane bend	E_u	183			
			170 (depol)	B_{2g}	in-plane bend
out-of-plane bend	A_{2u}	93			

Of the three infrared-active modes, two are bends (A_{2u}, E_u) and one corresponds to a Pt-Cl stretch (E_u). The latter should come at higher energy and so can be assigned with some confidence, however distinguishing the remaining two bending modes is less easy. An educated guess, and it is no more than that, would assign the band at 183 cm^{-1} as the in-plane mode, as it is close in energy to the Raman-active in-plane bend at 170 cm^{-1}.

> *SAQ 5.5 : Assign the spectrum of $[NO_3]^-$ based on your answers to SAQ 5.2 - SAQ 5.4.*

Infrared (cm^{-1}, solid)	Raman (cm^{-1}, solution)
1383	1385 (depol)
	1048 (pol)
825	
720	718 (depol)

In our analyses of the stretching and bending modes of $[PtCl_4]^{2-}$ we have made no comment on what these stretches and bends look like. We did do this for the simpler case of SO_2 in Chapter 3, where the two stretches – symmetric and anti-symmetric – and the one bending mode are self-evident. It is more difficult, but still possible, to evaluate the nature of the stretches and bends of more complex species such as $[PtCl_4]^{2-}$ using a technique known as **projection operators**. This is a non-trivial process for all but the simplest cases and is treated at a basic level in Appendix 1.

5.2 THE VIBRATIONAL SPECTRUM OF Xe(O)F$_4$

The final section of this chapter is a summary of all the key points made earlier in Part II, in the form of a complete worked example of how to fully analyse a vibrational spectrum. The molecule we will look at is Xe(O)F$_4$, whose vibrational spectrum is given in Table 5.2.

Table 5.2 The vibrational spectrum of Xe(O)F$_4$.

Infrared (vapour; cm^{-1})	Raman (liquid; cm^{-1})
926	920 (pol)
609	605 (depol)
576	567 (pol)
	527 (depol)
361	364 (depol)
288	286 (pol)
	232 (depol)
	220 (depol)
159	160 (depol)

The steps involved in assigning these data are highlighted by the following bullet points.

- determine the correct molecular shape and point group:

VSEPR predicts an octahedral arrangement of six electron pairs with one site occupied by a lone electron pair. The shape is a square-based pyramid. The molecule has a principal C_4 axis, no C_2 axes perpendicular to C_4 nor a σ_h. It does have vertical mirror planes, two of which are σ_v, the other two being σ_d. The point group is C_{4v}.

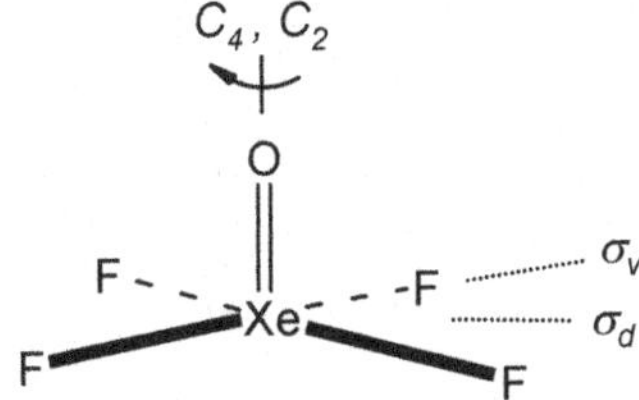

- determine Γ_{3N} using three vectors on each atom; to do this, you need to identify the number of unshifted atoms for each operation and multiply that number by $\chi_{\text{unshifted atom}}$:

	E	$2C_4$	C_2	$2\sigma_v$	$2\sigma_d$
unshifted atoms	6	2	2	4	2
$\times\ \chi_{u.a.}$	3	1	–1	1	1
Γ_{3N}	18	2	–2	4	2

- reduce Γ_{3N} to the sum of the irreducible representations of the point group:

$$A_1 = 1/8[(1 \times 18 \times 1) + (2 \times 2 \times 1) + (1 \times -2 \times 1) + (2 \times 4 \times 1) + (2 \times 2 \times 1) = 4$$
$$A_2 = 1/8[(1 \times 18 \times 1) + (2 \times 2 \times 1) + (1 \times -2 \times 1) + (2 \times 4 \times -1) + (2 \times 2 \times -1) = 1$$
$$B_1 = 1/8[(1 \times 18 \times 1) + (2 \times 2 \times -1) + (1 \times -2 \times 1) + (2 \times 4 \times 1) + (2 \times 2 \times -1) = 2$$
$$B_2 = 1/8[(1 \times 18 \times 1) + (2 \times 2 \times -1) + (1 \times -2 \times 1) + (2 \times 4 \times -1) + (2 \times 2 \times 1) = 1$$
$$E = 1/8[(1 \times 18 \times 2) + (2 \times 2 \times 0) + (1 \times -2 \times -2) + (2 \times 4 \times 0) + (2 \times 2 \times 0) = 5$$

$$\Gamma_{3N} = 4A_1 + A_2 + 2B_1 + B_2 + 5E$$

This set of symmetry labels adds up to 18 (note that E describes a doubly-degenerate mode), which is correct as it should correspond to 3N (3 x 6 = 18) modes.

- generate Γ_{vib} by removing from Γ_{3N} those labels which describe the molecular translations and rotations :

$$\Gamma_{trans + rot} = A_1 + A_2 + 2E \qquad (= 6, \text{correct})$$
$$\Gamma_{vib} = \Gamma_{3N} - \Gamma_{trans + rot}$$
$$\Gamma_{vib} = 3A_1 + 2B_1 + B_2 + 3E \qquad (= 3N\text{-}6 = 12, \text{correct})$$

- assign infrared and Raman activities to these modes:

A_1 : Infrared (**T**$_z$), Raman pol ($x^2 + y^2$, z^2).
B_1 : Raman depol ($x^2 - y^2$) .
B_2 : Raman depol (xy).
E : Infrared (**T**$_x$, **T**$_y$) and Raman depol (xz, yz).

It is possible to make some initial assignments at this point:

Infrared (vapour; cm^{-1})	Raman (liquid ; cm^{-1})	
926	920 (pol)	A_1
609	605 (depol)	E
576	567 (pol)	A_1
	527 (depol)	B_1 *or* B_2
361	364 (depol)	E
288	286 (pol)	A_1
	232 (depol)	B_1 *or* B_2
	220 (depol)	B_1 *or* B_2
159	160 (depol)	E

The A_1 and E modes are active in both the infrared and Raman, but only A_1 is Raman polarised; B_1 / B_2 are the only Raman-only modes.

- assign stretching modes using single-headed arrows to describe the Xe=O and Xe-F stretches:

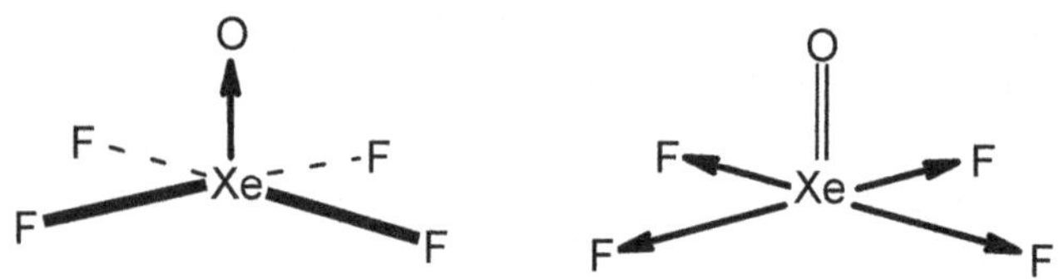

	E	$2C_4$	C_2	$2\sigma_v$	$2\sigma_d$	
$\Gamma_{Xe\text{-}O}$	1	1	1	1	1	$= A_1$
$\Gamma_{Xe\text{-}F}$	4	0	0	2	0	$= A_1 + B_1 + E$

- assign modes relating to deformations of the angles F-Xe-F and F-Xe-O:

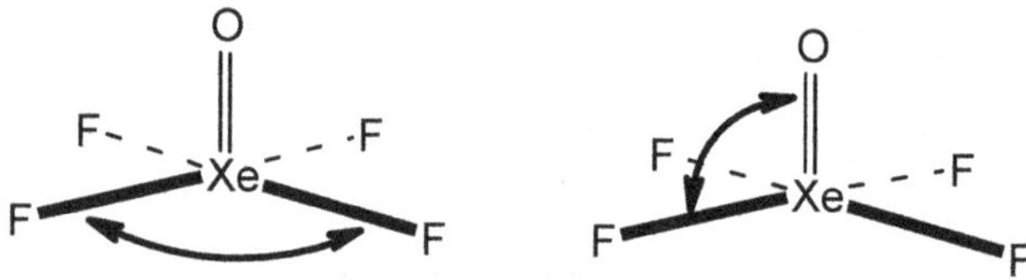

(only one of four of each of the F-Xe-F and F-Xe-O deformation vectors shown for clarity).

	E	$2C_4$	C_2	$2\sigma_v$	$2\sigma_d$	
$\Gamma_{F\text{-}Xe\text{-}F}$	4	0	0	0	2	$= A_1 + B_2 + E$
$\Gamma_{F\text{-}Xe\text{-}O}$	4	0	0	2	0	$= A_1 + B_1 + E$

One of the two A_1 bending modes must be redundant, as only $3A_1$ in total are allowed from Γ_{vib}. The redundancy relates to the F-Xe-F angular deformations, where all angles must add to 360°, which is not possible in the symmetric A_1 mode; this restriction does not apply to the four F-Xe-O angles.

- assign the spectrum:

Infrared (vapour; cm^{-1})	Raman (liquid ; cm^{-1})		
926	920 (pol)	A_1	Xe-O stretch
609	605 (depol)	E	Xe-F stretch
576	567 (pol)	A_1	Xe-F stretch
	527 (depol)	B_1	Xe-F stretch
361	364 (depol)	E	O-Xe-F *or* F-Xe-F deformation
288	286 (pol)	A_1	O-Xe-F deformation
	232 (depol)	B_1 *or* B_2	O-Xe-F *or* F-Xe-F defor'n
	220 (depol)	B_1 *or* B_2	O-Xe-F *or* F-Xe-F defor'n
159	160 (depol)	E	O-Xe-F *or* F-Xe-F deformation

Note the following:

- stretches come at higher energy than bends where a choice is possible e.g. the bands at 926 and 288 cm^{-1}.
- both Xe=O and Xe-F stretches have A_1 symmetry, though the Xe=O would be expected at higher energy as it is a stronger double bond.
- the two deformations of E symmetry cannot be distinguished.
- the two deformations of B_1 and B_2 symmetry cannot be distinguished.

5.3 GROUP FREQUENCIES

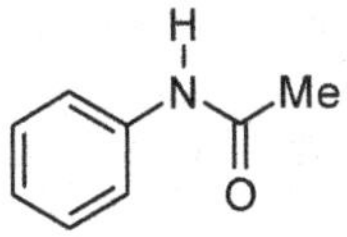

(1)

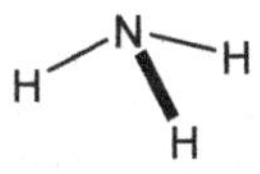

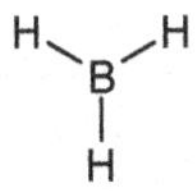

We began Chapter 1 by speculating on the way the N-H and B-H bonds might behave in each of the three species above and how group theory might help us resolve this problem. It was recognised at the outset that the symmetric species NH_3 and BH_3 would require an understanding of how symmetry restricts the behaviour of the three bonds involving hydrogen, whilst noting that the N-H bond in (**1**) was unique. As these molecules represent both high and low-symmetry species, after studying the way group theory can help with analysing the vibrational spectra of small symmetric molecules, it is pertinent to ask what the limitations are with regard to larger, more complex systems.

As molecules grow in complexity they become lower in symmetry – most of the many millions of known compounds will have C_1 symmetry i.e. no symmetry at all. The result of this is that each part of the molecule is unique and is not related to any other part, however chemically similar. As far as vibrational spectra are concerned, each bond can be considered in isolation, meaning that a spectrum becomes far too complex to analyse in its entirety. This is where a "group frequencies" approach comes into its own. Here, no attempt is made to analyse the spectrum completely, and only certain bands, those which are characteristic of key functional groups, are identified. Some typical group frequencies for a range of functionalities are given in Table 5.3, in addition to those already cited in Table 4.1.

Table 5.3 Typical infrared frequencies (cm^{-1}) for key functional groups.

Group	ν	Group	ν
C-H alkanes	2850 – 2960	O-H alcohols	3400 – 3640
C-H alkenes	3020 – 3100	O-H acids	2500 – 3100
C-H alkynes	3250 - 3340	B-H(terminal)	2450 – 2650
C=O aldehydes	1695 – 1740	B-H(bridging)	1600 – 2100
C=O ketones	1670 – 1725	P-H	2250 – 2450
C-O alcohols	1050 – 1150	S-H	2450 – 2650
C-O ethers	1085 – 1170	Si-H	2100 – 2250
C=S	1030 – 1080	Sn-H	1780 – 1910
S=O	1020 – 1060	Si-O	900 – 1200
C≡N nitriles	2210 – 2260	Si-F	600 – 1000
C=N	1630 – 1690	Si-Cl	400 – 750
C-N	1020 – 1220	Si-Br	250 – 600
NO_2	1545 – 1565	P-F	700 – 1050
N-H	3310 – 3500	P-Cl	400 – 600

In conclusion, when molecules are small and exhibit at least some symmetry, group theory can be a powerful aid in assigning vibrational spectra. However, as molecules become more complex and tend towards C_1 symmetry, any analysis will become limited to a group frequencies approach.

PROBLEMS

Answers to all problems marked with * are given in Appendix 4.

1. Determine Γ_{vib} and hence assign the spectrum of the perchlorate anion $[ClO_4]^-$. You may assume all Cl-O bonds are equivalent.

Infrared (solid; cm^{-1})	Raman (solution ; cm^{-1})
1111	1102 (depol)
	935 (pol)
625	628 (depol)
	462 (depol)

2*. Determine Γ_{vib} and hence assign the spectrum of $P(O)Cl_3$. Use six double-headed arrows to describe the six bond angle deformations.

Infrared (liquid; cm^{-1})	Raman (liquid ; cm^{-1})
1292	1290 (pol)
580	581 (depol)
487	486 (pol)
340	337 (depol)
267	267 (pol)
not accessible	193 (depol)

3*. Assign the vibrational spectrum of *trans*-difluordiazine, N_2F_2, given below:

Infrared (gas, cm^{-1})	Raman (gas, cm^{-1})
	1636 (pol)
	1010 (pol)
989	
	592 (pol)
412	
360	

Double-headed arrows are required for the two F-N-N in-plane bending modes and single-headed arrows on each fluorine describe the out-of-plane deformations.

4. In the gas phase MoF_5 adopts a trigonal bipyramidal geometry. Show that:

$$\Gamma_{vib} = 2A_1' + 2A_2'' + 3E' + E''$$

and hence assign, as fully as possible, the spectrum of this compound.

Infrared (gas; cm^{-1})	Raman (gas ; cm^{-1})
	747 (pol)
730	732 (depol)
	703 (pol)
685	
500	
	440 (depol)
240	239 (depol)
205	201 (depol)

(*Hint: by consideration of the analysis of SO_2 you should be able to identify the symmetric and anti-symmetric stretching modes involving the axial fluorines*).

5. $[Me_4N]Cl$ reacts with $InCl_3$ in a non-aqueous solvent to give $[Me_4N]_2[InCl_5]$. The vibrational spectrum of this species (region associated with the anion only) was recorded to determine whether the di-anion adopts a square-pyramidal (C_{4v}) or trigonal bipyramidal (D_{3h}) geometry :

Infrared (solid; cm^{-1})	Raman (solid ; cm^{-1})
294	294 (pol)
	287 (depol)
283	283 (depol)
274	274 (pol)
	193 (depol)
	165 (depol)
143	143 (depol)
140	140 (pol)
108	108 (depol)

Determine Γ_{vib} for both geometries and decide which structure is adopted. Assign the spectrum as fully as possible for the correct geometry.

6. The fluorosulphate anion $[FSO_3]^-$ can act in either a non-coordinating (**1a**) or bridging bidentate mode (**1b**) towards metals:

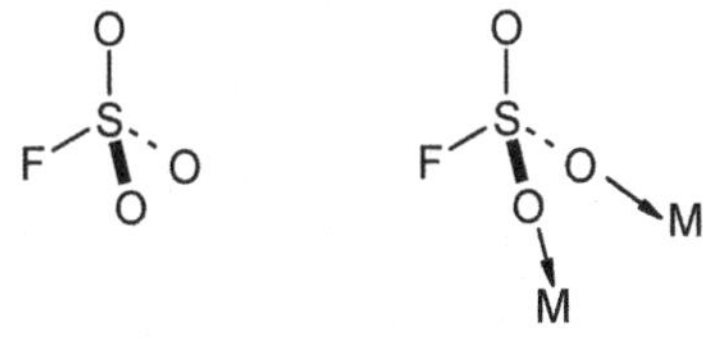

(1a) (1b)

Identify the point groups to which the $[FSO_3]^-$ ion belongs in each of the two situations.
Using appropriate S→O vectors, derive symmetry labels for the S-O stretching modes associated with each bonding type.
$K[FSO_3]$ exhibits infrared bands at 1287 and 1082 cm^{-1} associated with S-O stretches. What bonding mode is adopted by the anion in this species?

7. The Raman spectrum (cm^{-1}) of the $[SO_3]^{2-}$ ion contains bands at 966 (pol), 933 (depol), 620 (pol) and 473 (depol). Assign each of these bands, including the appropriate symmetry label.

8. OsO_4(pyridine) can adopt one of four structures (point groups treat pyridine as a single atom):

trigonal bipyramidal, axial N (C_{2v})
trigonal bipyramidal, axial O (C_{3v})
square pyramidal, axial N (C_{4v})
square pyramidal, axial O (C_s)

The vibrational spectral data (cm^{-1}) associated with the Os-O stretching region are:

Infrared	Raman
926	928 (pol)
915	916 (pol)
908	907 (pol)
885	886 (depol)

Determine the infrared and Raman activities for the Os-O stretches for each structure and hence determine which geometry is most consistent with the observed data.

PART III

APPLICATION OF GROUP THEORY TO STRUCTURE AND BONDING

6

FUNDAMENTALS OF MOLECULAR ORBITAL THEORY

In the opening chapter of this part of the book we review the fundamental concepts which underpin the construction of molecular orbital (MO) diagrams, which form the basis of our understanding of both molecular structure and reactivity and are an essential part of modern chemistry. The ideas presented, using the simple molecules H_2 and H_3, will form the basis on which we use group theory to guide us in developing descriptions of more complex bonding patterns. The power of group theory will be evident as we move from the interaction of the two or three atomic orbitals (AOs) of H_2, H_3 in this chapter to combinations of 19 AOs available for bonding in ferrocene, $(C_5H_5)_2Fe$ in Chapter 10, with which we shall conclude this part of the book.

6.1 BONDING IN H_2

When two hydrogen atoms approach each other to form a bond and construct a molecule of H_2, the two electrons, one on each of the two atoms, interact with each other. This interaction can be either in-phase and constructive, both electron waves having the same amplitude (ψ), or out-of-phase and destructive, in which the two electron waves have opposite amplitude. These possible combinations of the two 1*s* AOs generate two new MOs associated with the whole molecule, one bonding and one anti-bonding in character. The bonding MO concentrates electron density between the nuclei and is thus lower in energy than the AOs from which it is derived, while the anti-bonding MO has reduced electron density between the nuclei and is raised in energy relative to the AOs. In mathematical terms, these two new electron waves (ψ_{bonding} and $\psi_{\text{anti-bonding}}$) can be written in terms of the electron waves associated with each separate hydrogen atom (ψ_1, ψ_2) as follows:

$$\psi_{\text{bonding}} = \psi_1 + \psi_2 \qquad \text{(eqn. 6.1)}$$
$$\psi_{\text{anti-bonding}} = \psi_1 - \psi_2 \qquad \text{(eqn. 6.2)}$$

This situation can be represented pictorially in an MO diagram, which for H_2 is:

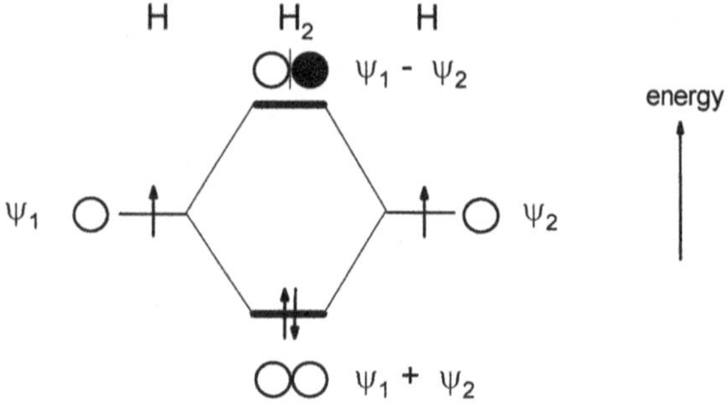

Fig. 6.1 MO diagram for H_2.

In Fig. 6.1, the open and filled 1*s* orbitals represent opposite phases for the electron waves. For the anti-bonding MO, there is a phase change at the mid-point of the H-H bond where $\psi = 0$, known as a **node**. Nodes occur where the electron density is zero between the two nuclei and thus correlate with an increase in energy: the two positively-charged nuclei see more of each other and are repelled, so weakening the bond. For H_2, the bonding MO, with no nodes, is lower in energy than the anti-bonding orbital which has one node.

Filling of these MOs follows both the Pauli exclusion principle and Hund's rule (which also apply to the filling of AOs), i.e. each MO can hold two electrons of opposite spin, and degenerate orbitals are filled separately with electrons of parallel spin until all these orbitals are half-filled.

Two general rules relating to the construction of MO diagrams emerge from this simple example:

- *n* AOs combine to generate *n* MOs.

In the case of H_2, two AOs generate two MOs; the total number of orbitals can neither increase nor decrease in the process of combination.

- the energy of an MO increases as the number of nodes increases.

6.2 BONDING IN LINEAR H_3

In the case of H_2 it is easy to appreciate how the two AOs combine, as in-phase and out-of-phase combinations are the only possibilities. How, then, do we approach the situation for H_3 involving three AOs ? Let us assume at this stage the molecule is linear, and return to the possibility that it is triangular in Chapter 8. Our strategy will be step-wise, and will consider the orbitals on the terminal atoms first and see how they can combine, then look at possible interactions between these terminal atom combinations and the 1*s* orbital on the central hydrogen. The two terminal hydrogen 1*s* AOs of H_3 can combine either in-phase or out-of-phase, as with H_2, above. However, only the in-phase combination of terminal atom AOs can interact with the central atom AO, with both bonding and anti-bonding possibilities:

The corresponding out-of-phase combination of terminal atom 1*s* AOs has equal amounts of in-phase (bonding) and out-of-phase (anti-bonding) interactions with the central atom AO and thus is non-bonding overall:

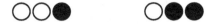

It does not matter what phase the central 1*s* AO has, it will generate equal amounts of a bonding interaction to one side and an anti-bonding interaction to the other. There is no net contribution from the central AO and it is ignored; the out-of-phase combination of terminal atom AOs thus generates a non-bonding MO.

The MO diagram for H_3 is shown in Fig. 6.2.

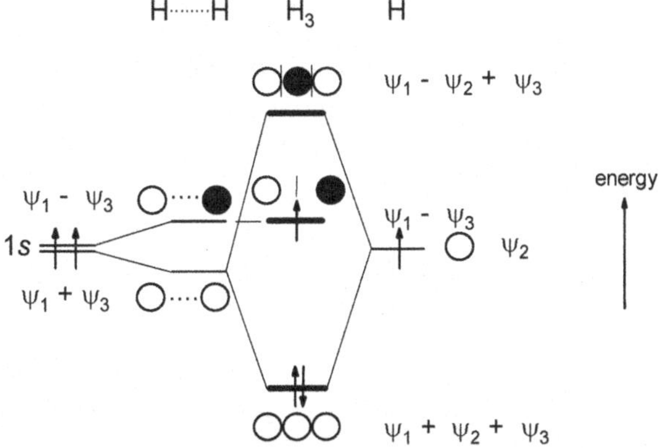

Fig. 6.2 MO diagram for linear H_3.

Note that we have created three MOs (bonding, non-bonding and anti-bonding) from three AOs, and that the energy of the three MOs increases as the number of nodes increases.

One final general observation can be made from the MOs in Fig. 6.2 :

- the nodes must be symmetrically positioned along the MO.

Thus, for the non-bonding MO of H_3 the node is centrally located, rather than offset from the middle.

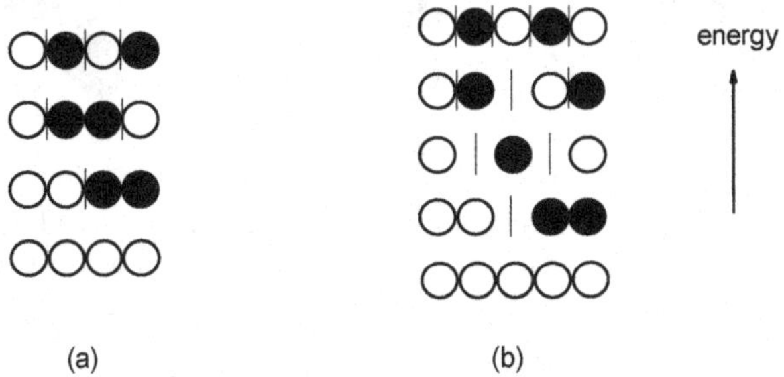

Fig. 6.3 Nodal arrays for (a) linear H_4 and (b) linear H_5.

This generalisation allows us to generate nodal sequences for the MOs associated with longer linear arrays, such as H_4, H_5 etc (Fig. 6.3). H_4 will have four combinations arising from the four AOs, and so on.

SAQ 6.1 : Draw the six MOs for linear H_6 showing the symmetric nodal arrangements.

Answers to all SAQs are given in Appendix 3.

These arrays will form the basis on which we combine the atomic orbitals belonging to a group of ligands surrounding a central atom, and their importance will become clearer in later chapters.

MO diagrams are clearly not limited to bonding schemes which include only hydrogen atoms or *s*-orbitals, as in the examples discussed above. The interactions of AOs can, however, be grouped into the type of bond that is formed, and this in turn depends on how the interacting AOs are oriented. When AOs approach in an "end-on" manner, the type of bond that is formed is a σ-bond. This embraces the *s-s* interactions in H_2 and H_3 but a more explicit example would be the "end-on" approach of two p_z orbitals (Fig. 6.4a):

Fig. 6.4 (a) "End-on" overlap of *p*-orbitals to generate a σ-bond and (b) the view of the MO down the bond axis.

Viewed along the bond (z) the orbital looks like an *s*-orbital i.e. it has cylindrical symmetry and no nodes. When a σ-bond is rotated around the bond axis there is no phase change (Fig. 6.4b). This defines a σ-bond.

When two *p*-orbitals approach "side-on", the bond that is formed is a π-bond (Fig. 6.5a); viewed along the direction of bond formation, the MO looks like a *p*-orbital and has one node (Fig. 6.5b). Thus, the side-on interaction of two *d*-orbitals also generates a π-bond (Fig. 6.5c)

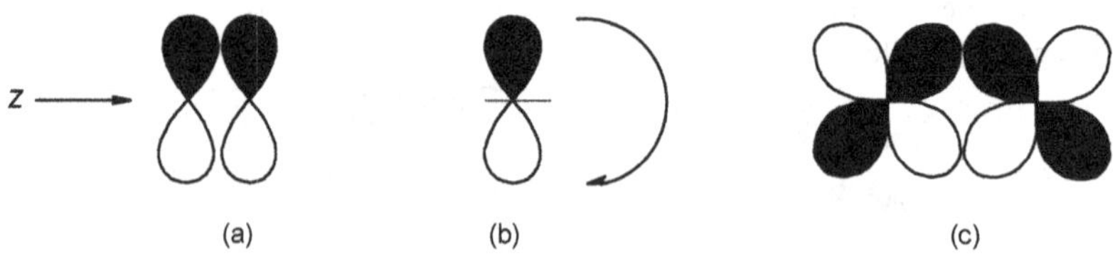

Fig. 6.5 (a) "Side-on" overlap of *p*-orbitals to generate a π-bond, (b) the view of the MO down the bond axis and (c) formation of a π-bond from side-on overlap of *d*-orbitals.

The "face-on" overlap of *d*-orbitals is an important interaction for transition elements and serves to generate a δ-bond (Fig. 6.6a); viewed along the direction of bond formation the MO looks like a *d*-orbital and has two nodes (Fig. 6.6b):

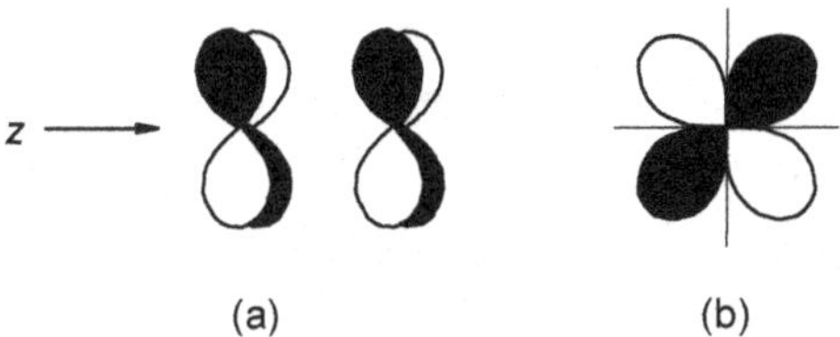

Fig. 6.6 (a) "Face-on" overlap of *d*-orbitals to generate a δ-bond and (b) the view of the MO down the bond axis.

The extent of AO overlap dictates the strength of the bond (or weakness of the anti-bond) that is generated. As the atoms come together, repulsions between nuclei and repulsions between non-bonding (i.e. core) electrons begin to set in, restricting how close the atoms can approach to each other. End-on overlap is thus more effective than side-on overlap, which in turn is more effective than any face-on interaction. Bond strengths between like atoms therefore follow the general order:

$$\sigma > \pi > \delta$$

> *SAQ 6.2 : Draw the AO combinations which describe the π-MOs in the allyl anion :*
>
> *Using Figure 6.2 as a guide, draw an MO diagram to determine the π-bond order in this anion.*
> (Hint: assume sp^2 hybridisation at each carbon)

6.3 LIMITATIONS IN A QUALITATIVE APPROACH

The way this book will approach the construction of MOs, using group theory as an aid, will be a very pictorial one. This qualitative approach, while avoiding some of the more daunting mathematical aspects of the subject, does have its limitations. These are apparent when the MO diagram of a heteronuclear diatomic such as HF is generated (Fig. 6.7, *overleaf*). Here, the bonding involves the $2p_z$ orbital on fluorine and the 1*s* orbital on hydrogen (the bond axis is conventionally taken as lying along *z*). The two atomic orbitals combine "end-on" to generate σ-bonding and anti-bonding combinations, with the $2p_x$, $2p_y$ AOs on fluorine being non-bonding.

Although the two AOs combine in much the way they did for H_2 (Fig. 6.1), there is a key difference : the energies of the two AOs are not the same. The ionisation energies are 13.6 and 18.7 eV for the H 1*s* and F $2p_z$, respectively, i.e. the fluorine 2*p* orbitals are at a much lower energy than the hydrogen 1*s*. The implications of this are two-fold:

- the extent of interaction between the AOs decreases as the difference in their energy increases.

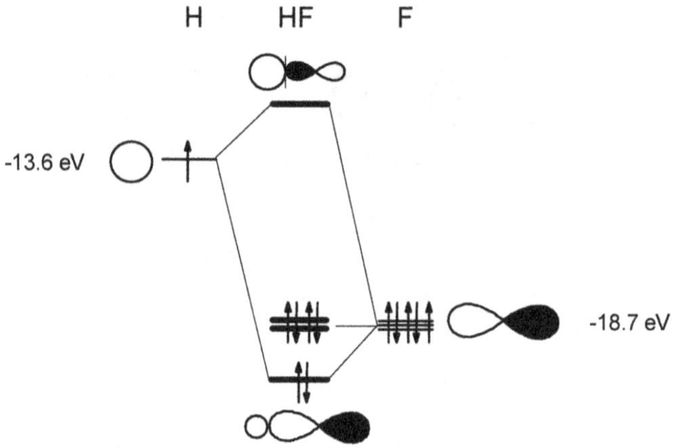

Fig. 6.7 An MO diagram for HF. The non-bonding $2p_x$, $2p_y$ AOs on fluorine are not shown for clarity.

The two 1*s* orbitals of H_2 interact strongly as they have the same energy, resulting in a significant stabilisation of the bonding MO and an equally large destabilisation of the anti-bonding MO. For HF, the extent of interaction is much smaller, so, for example, the bonding MO is only slightly lower in energy than the fluorine $2p_z$.

- the MOs have differing contributions from the participating AOs.

The bonding MO has a major contribution from the fluorine $2p_z$, while the anti-bonding MO is dominated by the contribution from the hydrogen 1*s*; the MOs appear much like the AO with which they are closest in energy. In effect, the lower energy AO (fluorine $2p_z$) has been slightly stabilised by an in-phase contribution from the hydrogen 1*s*, while the higher energy AO (hydrogen 1*s*) has been equally destabilised by its out-of-phase interaction with the fluorine AO. Pictorially, these effects are reflected in the size of the respective AOs in the diagrammatic representations of the two MOs in Fig. 6.7. The two electrons which form the bond, one from each atom, reside in an MO that is largely centred on fluorine, and this leads to a polar $H^{\delta+}$-$F^{\delta-}$ bond.

In more quantitative terms, the new electron waves associated with the two MOs are:

$$\psi_{\text{bonding}} = N[c_1\psi_1 + c_2\psi_2] \quad \text{(eqn. 6.3)}$$

$$\psi_{\text{anti-bonding}} = N[c_1\psi_1 - c_2\psi_2] \quad \text{(eqn. 6.4)}$$

N is a ***normalisation constant*** which is there to make sure the total electron density (ψ^2 summed over all space) is equal to one electron. More important are the coefficients c_1 and c_2, which are associated with the contributions of the two AOs to each MO. In the case of H_2, $c_1 = c_2$ since the two atoms are identical, while for HF $c_1 >> c_2$ for the bonding MO and *vice versa* for the anti-bonding case. This is a more quantitative expression of the information shown in the MO diagram of Fig. 6.7.

In presenting a qualitative, pictorially-based approach to MO construction we will ignore these coefficients, as a more rigorous approach requires both a knowledge of the relevant AO ionisation energies and also, importantly, a detailed computational analysis of how the waves interact. We will only comment on these coefficients when our qualitative approach demands it for clarity.

6.4 SUMMARY

- *n* AOs combine to generate *n* MOs.
- the energy of an MO increases as the number of nodes increases.
- the nodes must be symmetrically disposed along the MO.
- the interaction of AOs on terminal atoms is considered first and the resulting combinations combined, where possible, with the AOs on the central atom.
- σ-, π- and δ-MOs are formed by end-on, side-on and face-on overlap of AOs, respectively.
- for like atoms, the order of bond strength is $\sigma > \pi > \delta$.

PROBLEMS

Answers to all problems marked with * are given in Appendix 4.

1*. Predict the relative Xe-F bond strengths in $[XeF]^+$ and XeF_2 using suitable MO diagrams; consider only σ-bonds.

(*Hint* : *use* p_z *AOs on each atom*).

2. The infrared and Raman spectra (cm^{-1}) of $[HF_2]^-$, isolated in an argon matrix, are as follows:

Infrared	Raman
1550	
1200	
	675 (pol)

Draw an appropriate MO diagram consistent with the structure of the ion and hence determine the H-F bond order.

Is $[HF_2]^-$ diamagnetic or paramagnetic ?

3*. Draw an MO diagram for LiH and hence rationalise the bond polarity in this molecule.
(Valence orbital ionisation energies: Li 2*s* 5.4; H 1*s* 13.6 eV).

4*. Rationalise the C=O double bonds in CO_2 by constructing separate partial MO diagrams for (*i*) the σ-bonds (using 2*s* and $2p_z$ orbitals on carbon and $2p_z$ on oxygen) and (*ii*) the π-bonds (using $2p_x$, $2p_y$ on each atom).
(Valence orbital ionisation energies: C 2*s* 19.5, 2*p* 10.7; O 2*p* 15.9 eV).

(*Hint* : *none of the anti-bonding orbitals are occupied*).

7

H_2O – LINEAR OR ANGULAR ?

In constructing the MOs for H_2 we were faced with a simple problem, because there were only two possible ways in which the two AOs could interact – either in-phase or out-of-phase. When we move to a diatomic molecule which includes elements from the second row of the Periodic Table, e.g. O_2, the problem becomes significantly more complex. We need to consider all the possible combinations of electron waves between the two atoms i.e. between electrons in the 1*s*, 2*s* and each of the three 2*p* orbitals. There are two criteria which are important in evaluating the allowed electron-electron interactions:

- only electrons of similar energy interact significantly.

This restriction has been commented on already with regard to the bonding in HF (*Section 6.3*). In the case of O_2, the energy differences between the three different types of AO mean that only 1*s*-1*s*, 2*s*-2*s* and 2*p*-2*p* interactions, but not, for example 1*s*-2*s*, are energetically plausible. Secondly:

- only orbitals of the correct orientation can interact.

This means, for example, that a $2p_z$-$2p_z$ combination is allowed (Fig. 7.1a) and generates σ-bonding and σ-anti-bonding MOs, while $2p_z$-$2p_y$ overlap generates equal amounts of bonding and anti-bonding character and results in no change in energy (Fig. 7.1b).

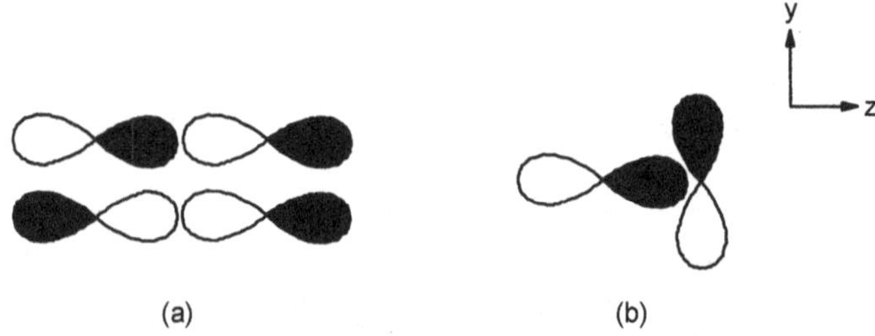

Fig. 7.1 (a) Allowed and (b) non-allowed combinations of 2*p* AOs.

In effect, we can re-phrase the second of the above criteria in terms of symmetry:

- only orbitals of the same symmetry can interact.

This chapter focuses on the role of group theory in defining the symmetries of the AOs associated with each atom in a molecule, how these orbitals can be grouped together and hence which combinations of AOs are allowed on bond formation. We will explore the bonding in H_2O by way of introducing the methodology.

7.1 SYMMETRY-ADAPTED LINEAR COMBINATIONS

We have already established the stepwise approach to generating an MO diagram in Section 6.2 :

- determine the way the AOs on the terminal atoms (hydrogen) combine.
- see how these combinations interact with the AOs on the central atom (oxygen).

In fact, we have already seen how the two hydrogen 1*s* orbitals combine in a pictorial fashion:

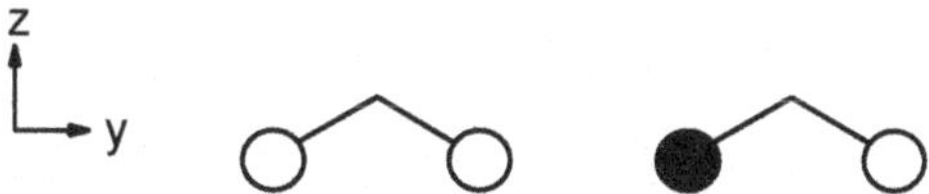

Fig. 7.2 Combinations of hydrogen 1*s* orbitals in H_2O.

As a prelude to more complex situations, we need to establish how group theory can bring us logically to this answer. Using the two hydrogen 1*s* AOs as a basis set we can generate the following reducible representation for the C_{2v} point group, using the same approach as previously for vectors (*Part II*).

C_{2v}	E	C_2	$\sigma(xz)$	$\sigma(yz)$
$\Gamma_{H\,1s}$	2	0	0	2

$= a_1 + b_2$

The two AOs do not move under E and $\sigma(yz)$ (count 1) while they move to new positions (in fact, they swap places) under C_2 and $\sigma(xz)$ (count 0). The reducible representation is easily converted to the sum of the a_1 and b_2 irreducible representations, either by inspection of the C_{2v} character table or by using the reduction formula (*Section 3.2*).

Group theory thus tells us that there are two ways in which the hydrogen 1*s* orbitals can be combined, and these combinations have a_1 and b_2 symmetries. Note that we have generated labels for *two* combinations, as expected given that *n* AOs combine to give *n* MOs (*Section 6.1*). By convention, lower case symmetry labels (Mulliken symbols) are used to describe orbitals, in contrast to the use of upper case labels to describe vibrational symmetry.

By considering the two possibilities shown in Fig. 7.2 as the basis sets, we can make the assignments:

C_{2v}	E	C_2	$\sigma(xz)$	$\sigma(yz)$	
$\Gamma_{\text{in-phase}}$	1	1	1	1	$= a_1$
$\Gamma_{\text{out-of-phase}}$	1	−1	−1	1	$= b_2$

Note that each combination of AOs must now be considered as a complete unit, as we did with the symmetric and anti-symmetric stretching modes of SO_2 (*Section 3.3*). Thus, under C_2 the out-of-phase combination reverses on itself and counts −1 to the representation:

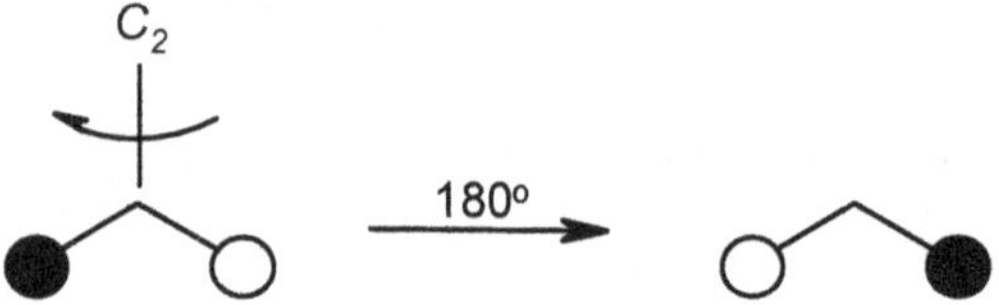

The two combinations of AOs shown in Fig.7.2 are termed **Symmetry-Adapted Linear Combinations** or **SALCs**. They are *linear* combinations as the two electron waves are either added to, or subtracted from, each other (eqns 6.1, 6.2) and have been *adapted* by considering the restrictions imposed by the *symmetry* of the C_{2v} point group.

7.2 CENTRAL ATOM ORBITAL SYMMETRIES

We now need to establish the symmetry labels which describe the AOs on the central oxygen atom. We need only consider the valence orbitals (2*s*, 2*p*) as the core 1*s* electrons are too tightly held by the nucleus to participate in bonding. Using each of the valence AOs in turn as basis set we have the following series of representations:

C_{2v}	E	C_2	$\sigma(xz)$	$\sigma(yz)$	
$\Gamma_{O\,2s}$	1	1	1	1	$= a_1$
$\Gamma_{O\,2px}$	1	−1	1	−1	$= b_1$
$\Gamma_{O\,2py}$	1	−1	−1	1	$= b_2$
$\Gamma_{O\,2pz}$	1	1	1	1	$= a_1$

Note that some orbitals reverse under certain symmetry operations (e.g. p_y under C_2) and count −1, as we have already noted in Section 2.3.

An important generalisation can be drawn from the above set of symmetry labels:

- the labels associated with the three *p*-orbitals are the same as those for the corresponding translational vectors:

$$p_x \equiv \mathbf{T_x} = b_1$$
$$p_y \equiv \mathbf{T_y} = b_2$$
$$p_z \equiv \mathbf{T_z} = a_1$$

The symmetries of the central atom *p*-orbitals can therefore be read directly from the character table, without further analysis. Indeed, the determination of central atom AO symmetries is only valid here because each AO is unique (has an *a* or *b* label). In general, this is not the case, so reading the symmetries directly from the character table is the simplest method for determing these AO labels.

Furthermore:

- *s*-orbitals are spherically symmetric and always have the symmetry described by the irreducible representation shown in the first row of the character table (a character of 1 under all operations).

> *SAQ 7.1* : *What are the symmetries of the s- and p-orbitals under D_{4h} symmetry ?*
> *See Appendix 5 for the character table for D_{4h}.*
>
> **Answers to all SAQs are given in Appendix 3.**

7.3 A MOLECULAR ORBITAL DIAGRAM FOR H_2O

We can now produce an MO diagram for water, showing the MOs and their relative energies, by combining terminal atom SALCs with central atom AOs of the same symmetry:

Table 7.1 Symmetry-allowed orbital combinations between hydrogen SALCs and oxygen AOs for angular (C_{2v}) H_2O.

SALC	AO	Label	MOs
	$2s$	a_1	
	$2p_x$	b_1	
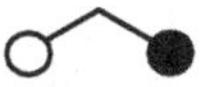	$2p_y$	b_2	
	$2p_z$	a_1	

Note how the nodal structure of each SALC matches that of the symmetry-related AO, so that areas of maximum (or minimum) electron density are aligned; this is particularly evident for the b_2 MO. For the two MOs of a_1 symmetry, the SALC has no nodes and neither does the *s*-orbital, or the lower lobe of p_z. On the other hand, the $2p_x$ AO on oxygen has no symmetry match with the SALCs on hydrogen and thus is non-bonding; the node of this AO cannot be matched with the a_1 SALC, which has no nodes, or the b_2 SALC, whose node is orthogonal to that of p_x. This alignement of nodes, which group theory delineates for us, is a feature of MO construction that you should look out for in subsequent examples.

In addition, the symmetrical SALC has the same a_1 symmetry as both $2s$ and $2p_z$ AOs on oxygen so, assuming all three orbitals have similar energy, they should be combined to generate three MOs. In the case of water, however, the relevant ionisation energies are H 1*s* 13.6, O 2*s* 32.4 and O 2*p* 15.9 eV, so the oxygen 2*s* is too low in energy to mix and is essentially non-bonding. The MO diagram is shown in Fig. 7.3; the two SALCs are shown with an average energy of –13.6 eV.

The numbers associated with the orbitals are present simply to distinguish MOs of the same symmetry label ($1a_1$, $2a_1$ etc); the two anti-bonding MOs have been identified with an asterisk (*) for clarity. The non-bonding $1a_1$ and $1b_1$ MOs are effectively the two lone pairs on oxygen. The total bond order is 2 (from the $2a_1$ and $1b_2$ MOs), which correlates with the two single O-H bonds.

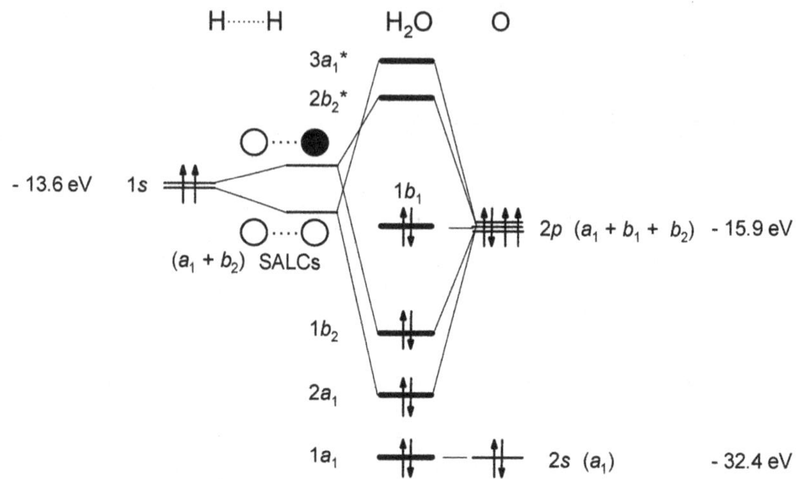

Fig. 7.3 MO diagram for angular (C_{2v}) H_2O.

7.4 A C_{2v} / $D_{\infty h}$ MO CORRELATION DIAGRAM

Any model of bonding must be able to rationalise the known properties of a molecule, such as molecular shape. How, then, can the MO description of water correlate with an angular, rather than a linear, geometry ? To answer this question we must compare the MO descriptions of water in each of the two cases.

The approach to generating an MO diagram for linear H_2O ($D_{\infty h}$) is the same as that for the C_{2v} structure, albeit that the character table for $D_{\infty h}$ looks a little daunting:

$D_{\infty h}$	E	$2C_\infty^\phi$....	$\infty\,\sigma_v$	i	$2S_\infty^\phi$....	∞C_2	
$\Gamma_{H\,1s}$	2	2	2	0	0	0	$= \sigma_g^+ + \sigma_u^+$

The "$2C_\infty^\phi$...." entry as a symmetry operation refers to the clockwise and anti-clockwise rotations about the C_∞ axis (lying along z) by any incremental angle ϕ; "$2S_\infty^\phi$...." has an analogous meaning.

The above reducible representation cannot be converted to the sum of irreducible representations using the reduction formula, as there are an infinite number of symmetry operations (g), so $1/g = 0$ (Section 3.2). The reduction can most easily be done by inspection, which is relatively easy in this case.

The symmetries of the s, p_x, p_y and p_z AOs can be read directly from the $D_{\infty h}$ character table (Appendix 5) and are σ_g^+ (top row), π_u ($\mathbf{T_x}$, $\mathbf{T_y}$) and σ_u^+ ($\mathbf{T_z}$), respectively. There are no matches for the $2p_x$, $2p_y$ AOs on oxygen, while the $2s$ and $2p_z$ AOs combine with the in-phase and out-of-phase SALCs, respectively, to give bonding and anti-bonding combinations.

Table 7.2 Symmetry-allowed orbital combinations between hydrogen SALCs and oxygen AOs for linear ($D_{\infty h}$) H_2O.

SALC	AO	Label	MOs
	$2s$	σ_g^+	
	$2p_x$	π_u	
	$2p_y$	π_u	
	$2p_z$	σ_u^+	

For consistency with Fig. 7.3, the oxygen $2s$ is assumed to be too low in energy to have a significant interaction with the σ_g^+ SALC.

The form of the MO diagram is:

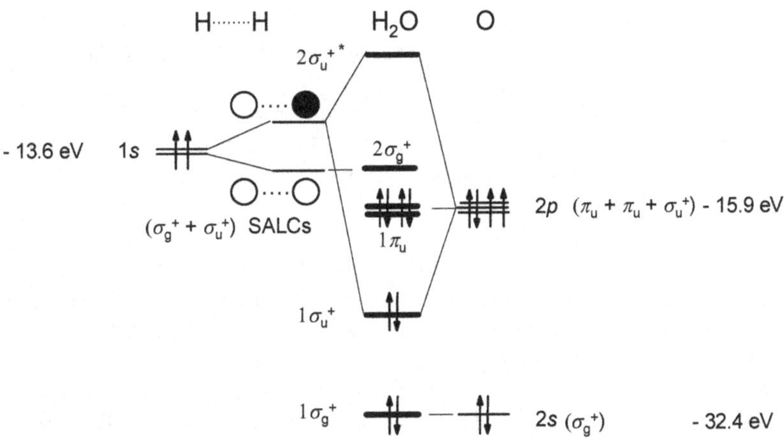

Fig. 7.4 MO diagram for linear ($D_{\infty h}$) H_2O.

There is clearly something wrong with the description of water as a linear molecule, as the O-H bond order implied by Fig. 7.4 is 0.5 !

The two MO descriptions are related by what is known as a correlation diagram (sometimes called a **Walsh correlation diagram**), which shows how the MOs describing the angular (C_{2v}) shape transform into the ones which describe the linear form ($D_{\infty h}$) as the H-O-H angle opens from 109° to 180°. This correlation diagram (Fig. 7.5) shows why H_2O adopts an angular shape with an O-H bond order of 1, rather than a linear structure with weaker O-H bonds.

All the possible interactions are shown for generality, ignoring any potential differences in energy between participating orbitals:

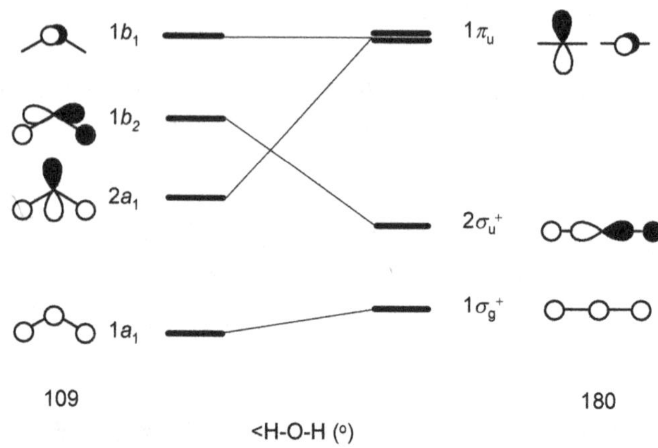

Fig. 7.5 MO correlation diagram for linear ($D_{\infty h}$) and angular (C_{2v}) H_2O.

In addition, no attempt has been made to calculate the magnitude of the energy changes involved, but we can make the following qualitative assessments:

$1a_1 \rightarrow 1\sigma_g^+$: the energy increases slightly as any terminal H···H bonding interaction is lost.

$2a_1 \rightarrow 1\pi_u$: rapid increase in energy as the MO goes from bonding to non-bonding.

$1b_2 \rightarrow 2\sigma_u^+$: energy decreases as (*i*) the end-on overlap of the *p*-orbital is more effective and (*ii*) any terminal H···H anti-bonding interaction is lost.

$1b_1 \rightarrow 1\pi_u$: no change in energy as the MO remains non-bonding in both geometries.

Clearly, the most significant changes involve the $2a_1$ and $1b_2$ MOs of C_{2v} symmetry. For water, the two bonding $2a_1$ electrons in the angular structure become non-bonding in the linear form (hence the reduction in O-H bond order), which is why H_2O prefers to retain an angular shape.

We are now in a position to rationalise the change in shape for a series of EH_2 (E = Be, B, C, N , O) species, based on filling of the MOs in Fig. 7.5. BeH_2 is predicted to be linear by VSEPR, and the four valence electrons (Be: 2e; 2 H: 2e) fill the two low energy σ^+ MOs of the linear arrangement. When additional valence electrons are involved (Table 7.3) retention of the linear shape would require the two high energy non-bonding π_u MOs to become occupied. To avoid this, the shape for all of these latter species distorts from linear to angular so that at least one of these MOs takes on a bonding character (a_1).

Table 7.3 Bond angles in EH_2.

	BeH_2	BH_2	CH_2	NH_2	OH_2
No. valence electrons	4	5	6	7	8
<H-E-H (°)	180	131	136	103	105

7.5 SUMMARY

- only orbitals of the same symmetry can interact.
- irreducible representations (symmetry labels) for the terminal atom SALCs are generated using group theory by counting 1, 0 or –1 as the contribution to the reducible representation, depending on if an AO is unmoved, moved or reversed, respectively, by a given symmetry operation.
- *s*-orbitals always have the symmetry of the perfectly symmetrical irreducible representation (1 for all operations).
- the symmetry labels associated with the three *p*-orbitals are the same as those for the corresponding $\mathbf{T_x}$, $\mathbf{T_y}$ or $\mathbf{T_z}$.
- MOs are generated by combining terminal atom SALCs with central atom AOs bearing the same symmetry label.
- orbital correlation diagrams showing the relative energy changes as two related shapes interconvert are helpful in predicting molecular geometry.

PROBLEMS

Answers to all problems marked with * are given in Appendix 4.

1*. Draw a partial MO diagram to describe the π-bonding in the nitrite ion $[NO_2]^-$ (C_{2v}), assigning a symmetry label to each MO:

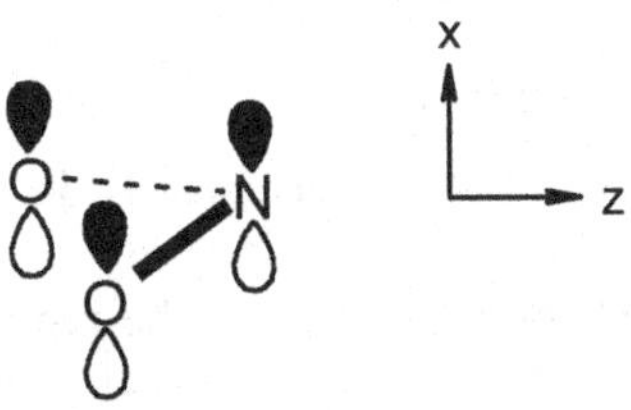

(Valence orbital ionisation energies : N 2*p* 13.1; O 2*p* 15.9 eV).
(*Hint*: *each oxygen has two lone pairs, nitrogen has one lone pair and both contribute one electron to the N-O σ-bond*).

2. Using the orbital correlation diagram for C_{2v} / $D_{\infty h}$, suggest why the first excited state of BeH_2 is angular while that of BH_2 is linear.

3. Adapt Fig. 7.3 to generate an MO diagram for angular (C_{2v}) BeH_2.
(Valence orbital ionisation energies : H 1*s* 13.6; Be 2*s* 9.3, 2*p* 6.0 eV).

8

NH_3 – PLANAR OR PYRAMIDAL ?

In this chapter we shall continue to refine our knowledge of how group theory can be applied to aid our understanding of bonding models, by considering the structures of BH_3 and NH_3. Can our MO descriptions rationalise why borane is planar while ammonia is pyramidal, in the same way that we rationalised the shapes of various EH_2 species in the Section 7.4 ? The approach outlined in Chapter 7 runs into a problem with EH_3 species, because the terminal atoms, the three hydrogens, adopt a cyclic (triangular) arrangement about the central atom, rather than the linear one which described the two terminal atoms in EH_2. Hence, we must first determine how to handle cyclic arrays of terminal atoms, and our vehicle for doing this will be a comparison of linear and triangular forms of H_3. Using the principles learnt from this exercise, we shall address the question of which structure is adopted by each of BH_3 and NH_3, before ending with a brief look at molecules with larger cyclic arrangements of terminal atoms.

8.1 LINEAR OR TRIANGULAR H_3 ?

We have already discussed the MOs for linear H_3 in Section 6.2, but it is easy to verify the diagram by applying the techniques of group theory outlined in the previous chapter. The point group is $D_{\infty h}$ and the representations for the terminal atom SALCs have already been determined as part of our analysis of linear H_2O. Thus, the symmetry of the terminal atom SALCs and the central atom 1*s* AO, are :

Terminal hydrogen SALCs : $\sigma_g^+ + \sigma_u^+$ Central hydrogen 1*s* : σ_g^+

The MO diagram for linear H_3 is shown in Figure 8.1 (*overleaf*), which is Figure 6.2 reproduced but with the addition of the symmetry labels determined by group theory - we have moved from constructing MOs by intuition to a more exact, more formalised symmetry-based approach.

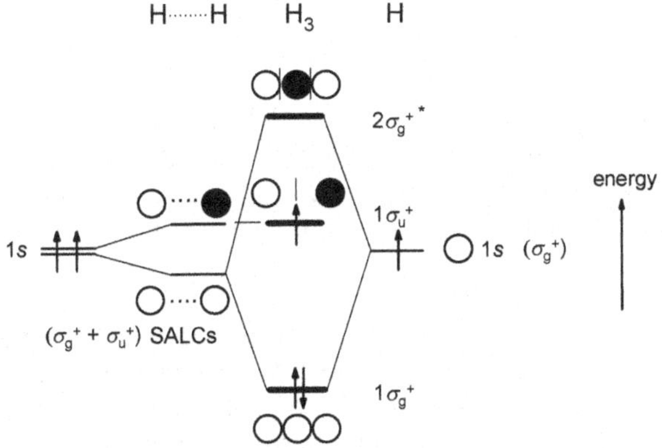

Fig. 8.1 MO diagram for H_3, including symmetry labels for the molecular orbitals.

For *triangular* H_3 the point group changes to D_{3h}. There is no longer a "central atom" but we can still treat all three hydrogens as a group, for which the representation is:

D_{3h}	E	$2C_3$	$3C_2$	σ_h	$2S_3$	$3\sigma_v$	
$\Gamma_{H\,1s}$	3	0	1	3	0	1	$= a_1' + e'$

We would expect three SALCs from combining three AOs, and the labels we have derived are consistent with this. What do these SALCs look like ? We can generate pictures of these SALCs by bending the SALCs for linear H_3 into a triangle:

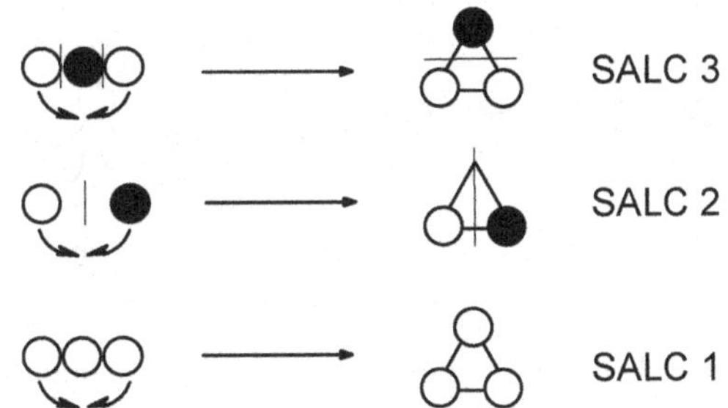

Fig. 8.2 SALCs for the hydrogen 1*s* orbitals in triangular H_3, derived from the SALCs for linear H_3.

It is easy to show that the lowest energy SALC – the one with no nodes – has a_1' symmetry, by counting 1, –1 or 0 if the SALC is unmoved, reversed or moved under a symmetry operation:

D_{3h}	E	$2C_3$	$3C_2$	σ_h	$2S_3$	$3\sigma_v$	
$\Gamma_{SALC\,1}$	1	1	1	1	1	1	$= a_1'$

The symmetry label for SALCs 2 and 3 cannot be rationalised in the same way as they have to be treated as a pair. This is a problem we have already encountered elsewhere, e.g. the $\mathbf{T}_x$ and $\mathbf{T}_y$ translational vectors under C_{3v} symmetry (Section 2.3). In general:

- only SALCs with either an *a* or a *b* label can have their symmetry labels confirmed by counting 1, –1 or 0 if the SALC is unmoved, reversed or moved under a symmetry operation.
- SALCs which are part of *e* or *t* degenerate sets cannot be analysed in this way.

The first point to note about SALCs 2 and 3 is that, despite having different numbers of nodes in the linear array, both these SALCs have one node in the triangular arrangement; this is because the two nodes of SALC 3 coalesce on bending. The fact that both SALCs have one node implies they are similar in energy, so an *e* label is reasonable. Note, however, that in general, SALCs with the same number of nodes do not necessarily have to be degenerate; if they are not described by *e* or *t* labels they are only similar, but not equal, in energy.

Since we now have MO diagrams for both linear and triangular H_3, we can draw a correlation diagram and use it to attempt to rationalise which of the two structures is preferred.

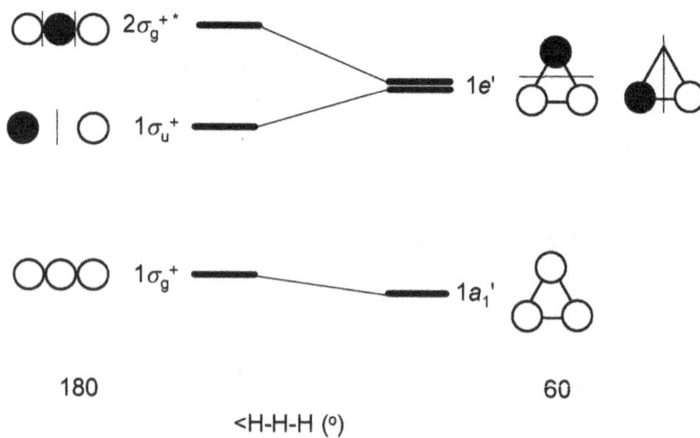

Fig. 8.3 MO correlation diagram for linear ($D_{\infty h}$) and triangular (D_{3h}) H_3.

As with the C_{2v} / $D_{\infty h}$ correlation diagram for water, we can draw some qualitative conclusions from the MO correlation diagram for H_3 in Fig. 8.3:

$1\sigma_g^+ \rightarrow 1a_1'$: the energy decreases with onset of H···H bonding at the base of the triangle.

$1\sigma_u^+ \rightarrow 1e'$: increase in energy as the MO goes from non-bonding to weakly anti-bonding.

$2\sigma_g^+ \rightarrow 1e'$: energy decreases as the MO goes from strongly anti-bonding to weakly anti-bonding (some bonding character is introduced at the base of the triangle).

It is not clear simply from this correlation diagram which geometry is favoured by a species with three valence electrons; calculations are required to determine the magnitude of the energy changes depicted before a conclusion can be reached. However, $[H_3]^+$, with only two valence electrons, would clearly favour the triangular shape as the electrons only occupy the lowest energy MO ($1a_1'$).

8.2 A MOLECULAR ORBITAL DIAGRAM FOR BH_3

We will begin by determining the MOs which describe planar BH_3 (D_{3h}). We already know the labels and appearance of the three SALCs associated with the hydrogen atoms, as they are the same as those derived for triangular H_3, which also has D_{3h} symmetry. The symmetries of the AOs on boron can be read directly from the character table (ignoring the core 1*s* AO):

2*s*	:	a_1' (*s*-orbitals are perfectly symmetrical)
$2p_x$, $2p_y$	:	e' (T_x, T_y)
$2p_z$	:	a_2'' (T_z)

Matches between the SALCs and the central atom AOs of identical symmetry are shown in Table 8.1.

Table 8.1 Symmetry-allowed orbital combinations between hydrogen SALCs and boron AOs for planar (D_{3h}) BH_3.

SALC	AO	Label	MOs
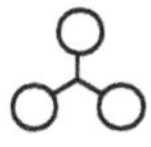	2*s*	a_1'	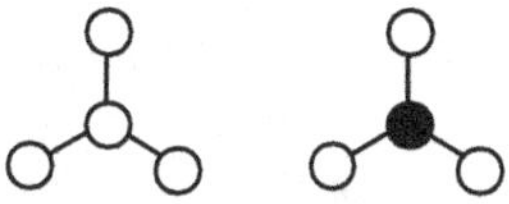
	$2p_x$	*e'*	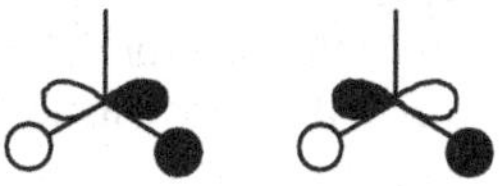
	$2p_y$	*e'*	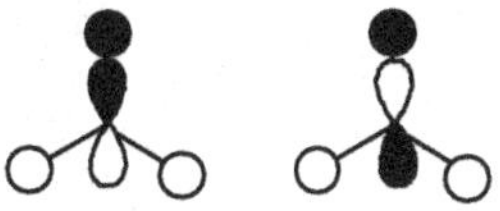
	$2p_z$	a_2''	

The 2*s*, $2p_x$ and $2p_y$ AOs all have symmetry matches with the terminal atom SALCs and pair up to generate bonding and anti-bonding MOs. The $2p_z$ AO has no symmetry match and is non-bonding.

The MO diagram for planar BH_3 is shown in Fig. 8.4 (*overleaf*). The B-H bond order is 3 in total i.e. each B-H bond has a bond order of 1, as we would expect.

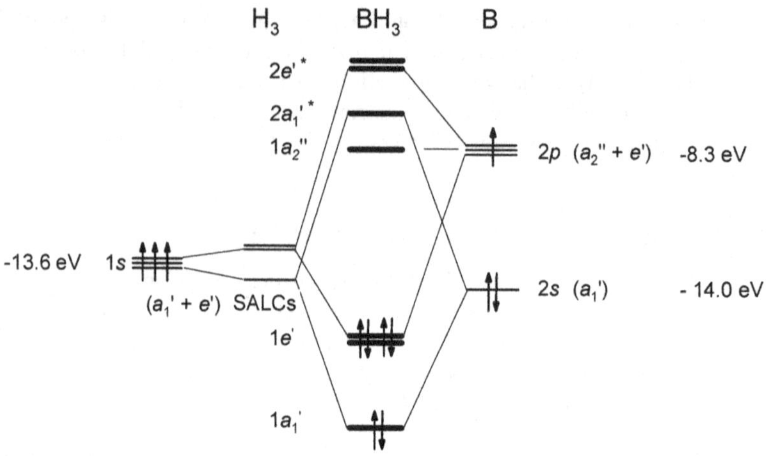

Fig. 8.4 MO diagram for planar (D_{3h}) BH_3.

The following SAQ will allow you to practise the application of the above techniques and derive descriptions of the MOs for pyramidal (C_{3v}) NH_3.

> *SAQ 8.1 : For NH_3, derive the symmetry labels for the H_3 SALCs and the nitrogen AOs under C_{3v} symmetry.*
> *Sketch each SALC and assign its symmetry label by consideration of the nodal arrays for cyclic H_3 (Fig. 8.2).*
> *Sketch the MOs generated by pairing SALCs and AOs of the same symmetry.*
>
> **Answers to all SAQs are given in Appendix 3.**

The MO diagram for the known pyramidal (C_{3v}) structure of NH_3 is shown in Fig. 8.5. The ordering of MOs is based on calculations, which indicate that, for this geometry, $2p_x$, $2p_y$ interact more effectively with the e SALCs than $2p_z$ does with the a_1 SALC, using as it does only its lower lobe.

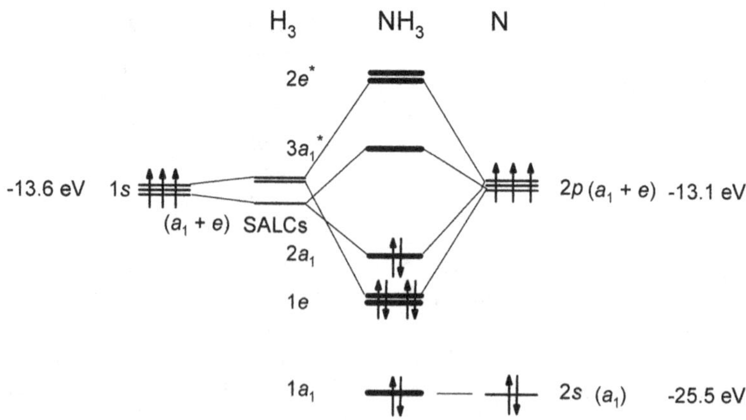

Fig. 8.5 MO diagram for pyramidal (C_{3v}) NH_3.

We can now generate a generalised orbital correlation diagram for EH_3 (Fig. 8.6) showing the relative energy changes as pyramidal C_{3v} distorts toward planar D_{3h}.

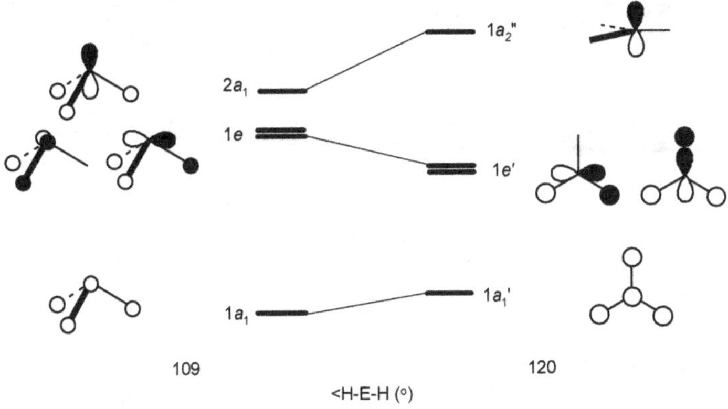

Fig. 8.6 MO correlation diagram for pyramidal (C_{3v}) and planar (D_{3h}) EH_3.

$1a_1 \rightarrow 1a_1'$: the energy increases slightly as any H···H bonding interaction across the base of the pyramid is lost.

$1e \rightarrow 1e'$: energy decreases slightly as (*i*) the in-plane overlap of the *p*-orbital with the SALC is more effective and (*ii*) any H···H anti-bonding interaction is reduced as the hydrogens move apart.

$2a_1 \rightarrow 1a_2''$: rapid increase in energy as the MO goes from bonding to non-bonding.

The significant difference between Figs. 8.4 and 8.5, highlighted in Fig. 8.6, is that the p_z AO on E in pyramidal EH_3 has a matching SALC and is thus bonding in character. Planar BH_3, with six valence electrons, just fills the a_1' and e' MOs of the D_{3h} structure; the empty a_2'' does not influence the shape, which is dictated by the lower energy of the e' over the e orbitals. On the other hand, NH_3 has eight valence electrons and a planar structure would place two of these in the high-energy, non-bonding a_2'' MO. Therefore, it opts for the pyramidal arrangement in which this latter MO becomes bonding ($2a_1$).

8.3 OTHER CYCLIC ARRAYS

The technique for producing cyclic nodal arrays from linear ones can be generalised and is the basis for descriptions of the bonding in molecules with four or more ligands around the central atom, in the same way as we have seen above for EH_3. The nodal arrays for cyclic H_5 and H_6 are shown in Figs. 8.7 and 8.8, while *SAQ 8.2* asks you to repeat the exercise for H_4. As with the example for H_3 (*Section 8.1*), some nodes coalesce on wrap-around.

The only other key rule that must be obeyed in carrying out this process is:

- all nodes must pass through the centre of the polygon.

This is in essence the same rule as we had for the arrangement of nodes for a linear array, where we stated that "*the nodes must be symmetrically disposed along the*

MO", applied to cyclic systems. In certain cases, this results in the movement of the nodes off atoms and three such cases are highlighted in Fig. 8.7 and Fig. 8.8.

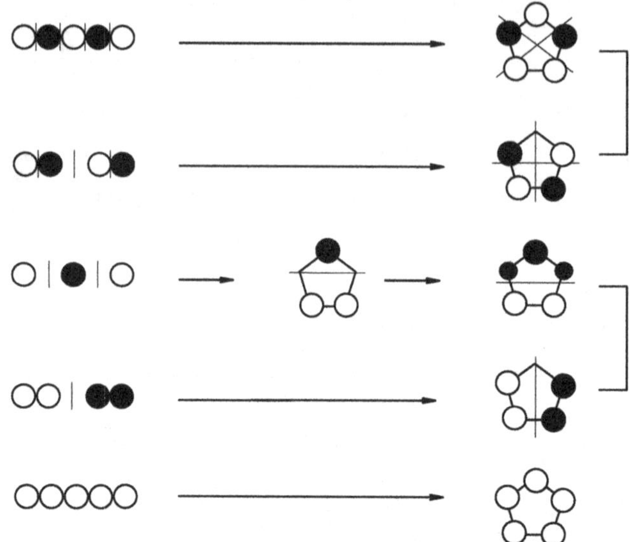

Fig. 8.7 Nodal patterns for cyclic H_5 derived from their linear analogues.

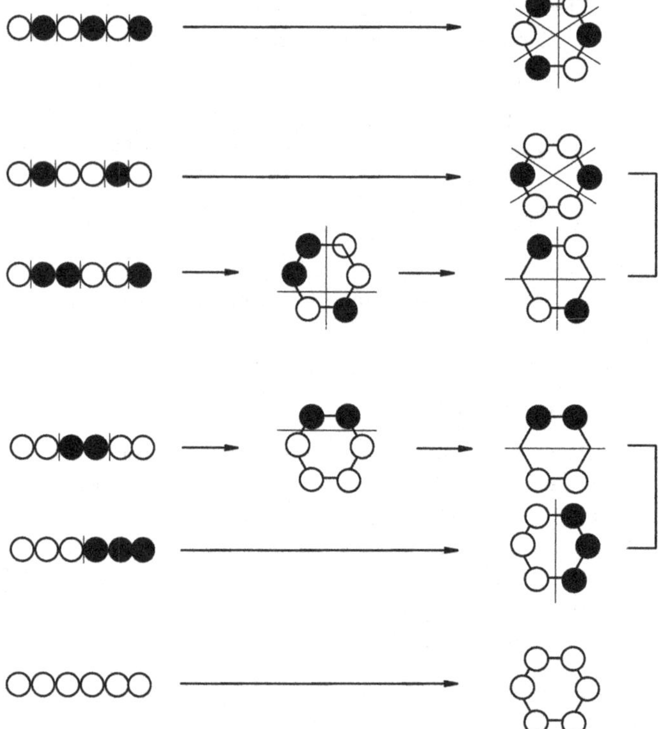

Fig. 8.8 Nodal patterns for cyclic H_6 derived from their linear analogues.

SAQ 8.2: *By consideration of the nodal patterns for linear H_4 (Fig. 6.3), draw the nodal pattern for the analogous cyclic system.*

As an example of how we can use these patterns to generate SALCs for more complex systems, let us consider the bonding in diborane, B_2H_6. This is often described as "*electron-deficient*" as the bridging B-H-B units only have two electrons in total, whereas, in general, we associate two electrons with any bond between two atoms i.e. four electrons would be expected for a B-H-B fragment. However, this idea of electron-deficiency, which implies marked chemical instability, is inconsistent with the known properties of B_2H_6, which is a stable, though air-sensitive, gas which is widely used as a synthetic precursor in boron chemistry.

Can we rationalise the stability of diborane with a bonding model ? How valid is the term "*electron-deficient*" for this molecule ? We will only focus on the construction of the two bridging B-H-B fragments and, for simplicity, will assume sp^3 hybridisation at boron. The terminal B-H bonds are formed by overlap of sp^3 hybrids with the $1s$ orbitals on hydrogen, generating conventional two-centre / two-electron σ-bonds.

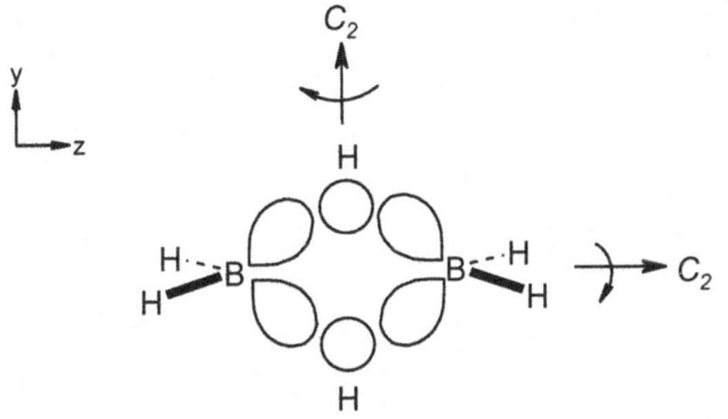

The first task is to identify the symmetry labels for the SALCs associated with the two hydrogens and with the four sp^3 hybrids, under point group D_{2h} symmetry:

D_{2h}	E	$C_2(z)$	$C_2(y)$	$C_2(x)$	i	$\sigma(xy)$	$\sigma(xz)$	$\sigma(yz)$	
$\Gamma_{H\,1s}$	2	0	2	0	0	2	0	2	$= a_g + b_{2u}$
$\Gamma_{B\,sp^3}$	4	0	0	0	0	0	0	4	$= a_g + b_{1u} + b_{2u} + b_{3g}$

The SALCs associated with the two hydrogens are the familiar in-phase (a_g) and out-of-phase (b_{2u}) combinations. The symmetry labels, specifically the subscripts g and u (symmetric and anti-symmetric with respect to inversion, respectively), differentiate these two SALCs and also allow us to go some way in assigning labels to the four SALCs associated with the boron sp^3 bridging orbitals (*see SAQ 8.2*):

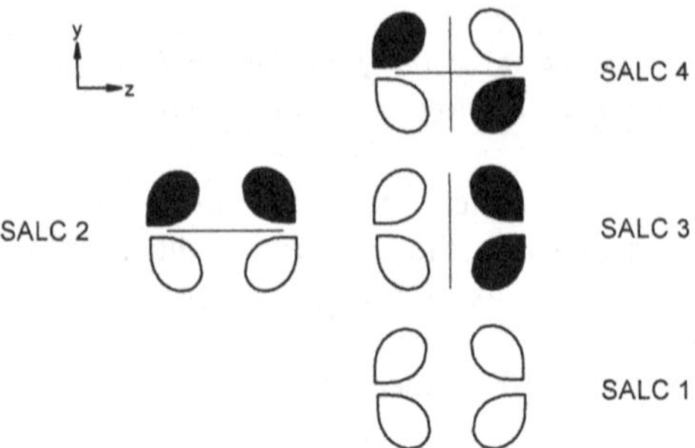

The two *g* SALCs are clearly identifiable (SALCs 1 and 4), the totally symmetric one having the *a* label (SALC 1). It is, however possible, given that all four of these SALCs have *a* or *b* labels, to identify them unambiguously:

D_{2h}	E	$C_2(z)$	$C_2(y)$	$C_2(x)$	i	$\sigma(xy)$	$\sigma(xz)$	$\sigma(yz)$	
$\Gamma_{SALC\ 1}$	1	1	1	1	1	1	1	1	$= a_g$
$\Gamma_{SALC\ 2}$	1	−1	1	−1	−1	1	−1	1	$= b_{2u}$
$\Gamma_{SALC\ 3}$	1	1	−1	−1	−1	−1	1	1	$= b_{1u}$
$\Gamma_{SALC\ 4}$	1	−1	−1	1	1	−1	−1	1	$= b_{3g}$

The MO diagram for the bridging units of B_2H_6 is then produced by linking SALCs with the same symmetry label.

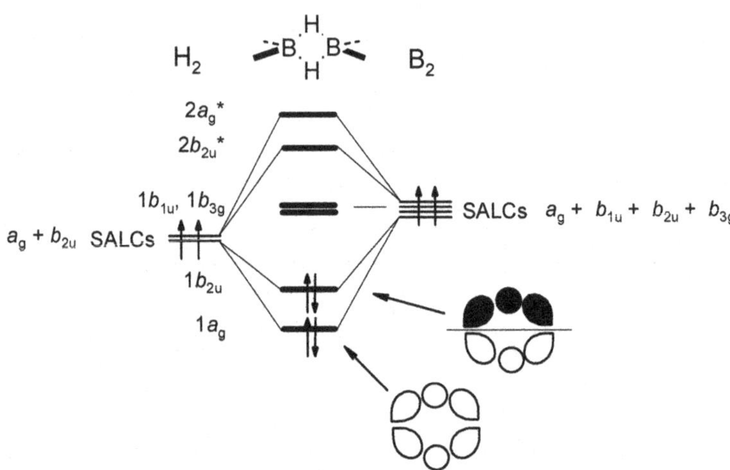

Fig. 8.9 MO diagram for the bridging B-H-B units of B_2H_6.

The $1a_g$ bonding MO is lower in energy than b_{2u} as the former has one less node. The four electrons available to support the two bridging fragments just fill the two available bonding MOs. On this basis the molecule is *not* electron-deficient, though the bond order for each B-H unit is only 0.5.

8.4 SUMMARY

- nodal patterns for cyclic arrays are generated from their linear analogues by bringing the ends of the linear arrays together.
- some nodes in the linear arrays coalesce on cyclisation.
- all nodes must pass through the centre of the polygon.
- orbitals with the same number of nodes distributed over the same number of atoms will have similar energies.
- only orbitals associated with either *e* or *t* labels are degenerate in energy.

PROBLEMS

Answers to all problems marked with * are given in Appendix 4.

1*. Generate an MO diagram to describe the π-bonding in BF_3.

What is the B-F π-bond order and why is BF_3 a weaker Lewis acid than might be expected ?

(Valence orbital ionisation energies : B $2p_z$ 8.3; F $2p_z$ 18.7 eV).

(*Hint*: *what is the hybridisation at boron in planar* BF_3 *? How are π-bonds being formed ?*)

2. Cyclobutadiene adopts a structure with two localised double bonds (D_{2h}; let the molecule lie in the *xy* plane) rather than the perfectly delocalised D_{4h} arrangement (*see* Fig. 5.1 for the location of the symmetry elements).

Determine the symmetry labels and orbital occupancy of the MOs responsible for the π-bonding in each of the two cases, and hence rationalise the observation stated above.

3. Based on the orbital correlation diagram for C_{3v} / D_{3h}, what structures do you predict for $[CH_3]^+$, $[CH_3]^\bullet$ and $[CH_3]^-$?

4. Draw MO diagrams for planar (D_{4h}) and tetrahedral (T_d) CH_4. To do this, follow the following steps for each geometry:

- determine the irreducible representations which describe the four hydrogen SALCs and the 2*s*, 2*p* AOs on carbon.

- draw the four SALCs and hence the way they interact with the carbon AOs.

  (*Hint : the SALCs for tetrahedral H_4 can be derived from those for planar H_4 by bending two opposite corners of the square up and the other two down*) :

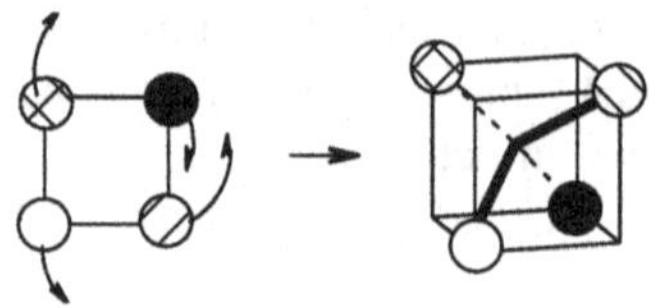

- arrange the MOs in order of increasing energy.

5. Using the MO sketches from question 5, draw an orbital correlation diagram for planar (D_{4h}) and tetrahedral (T_d) CH_4 and hence suggest why the tetrahedral form is adopted.

6*. Using the six p_z orbitals as basis set, determine the symmetry labels for the six π-MOs for benzene.

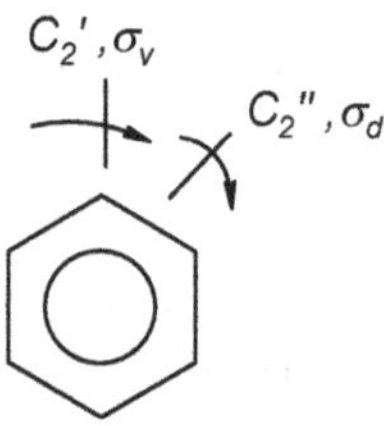

Adapt Fig. 8.8 to draw each of these MOs and assign each with the appropriate label.

9

OCTAHEDRAL COMPLEXES

Complexes in which the central atom has a coordination number of six are very commonly encountered for the majority of elements in the Periodic Table. These complexes are almost invariably octahedral in shape, the trigonal prism being a rare alternative, e.g. WMe_6.

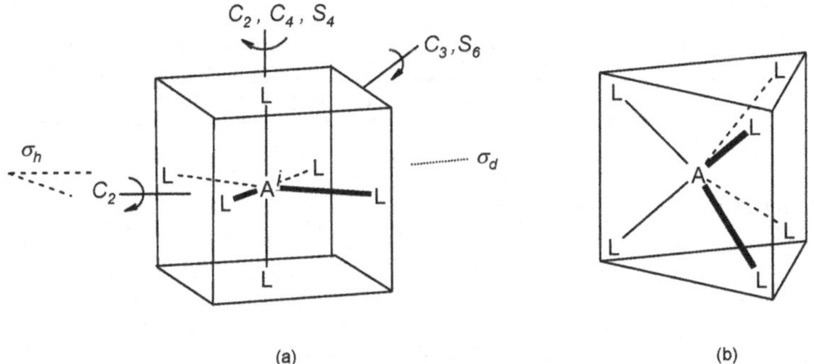

Fig. 9.1 Alternative shapes for six-coordinate AL_6 complexes: (a) octahedral, showing selected symmetry elements and (b) trigonal prism.

In this chapter we will apply the basic techniques developed in preceding chapters to rationalise some of the key features of octahedral complexes, i.e. the role of *d*-orbitals in the bonding of main group element compounds and the electronic structure of transition metal complexes. The introduction of transition elements into our bonding problems will require a knowledge of how group theory deals with the symmetry of *d*-orbitals and this is addressed in Section 9.2.

Many of the properties of transition metal species can be explained by simple crystal field theory (CFT), in which orbital energies – and hence their *d*-electron occupation – are determined by considering the relative spatial orientation of the *d*-orbitals and the ligands i.e. electrostatic repulsions between negatively charged ligands and the metal-based M^{n+} electrons are the key factors. This model is clearly a gross simplification, as octahedral species such as $[CoCl_6]^{3-}$ do not dissociate in solution into Co^{3+} and Cl^-, behaviour typical of ionic bonds. Moreover, the model becomes even less plausible when applied to organometallic systems such as $Cr(CO)_6$, which incorporate neutral ligands. Nonetheless, the majority of the conclusions reached using CFT are correct and as such it provides a useful starting point for understanding transition metal chemistry. There are, though, some key failings of CFT - not surprising given the simplicity of the model - which are catered for in a more sophisticated MO approach to bonding known as ligand field theory. We will review what this entails in the final part of this chapter.

9.1 SALCS FOR OCTAHEDRAL COMPLEXES

To begin with, let us only consider ligands which are σ-bonded to the central atom. Using *s*-orbitals to represent any σ-bonded ligand, our first task is to identify the symmetry labels for the six ligand SALCs under O_h symmetry (see Fig. 9.1 for the location of the key symmetry elements):

O_h	E	$8C_3$	$6C_2$	$6C_4$	$3C_2$ [a]	i	$6S_4$	$8S_6$	$3\sigma_h$	$6\sigma_d$
$\Gamma_{6L\sigma}$	6	0	0	2	2	0	0	0	4	2

[a] $C_2 \equiv C_4^2$.

This reduces to the following sum:

$$\Gamma_{6L\sigma} = a_{1g} + e_g + t_{1u}$$

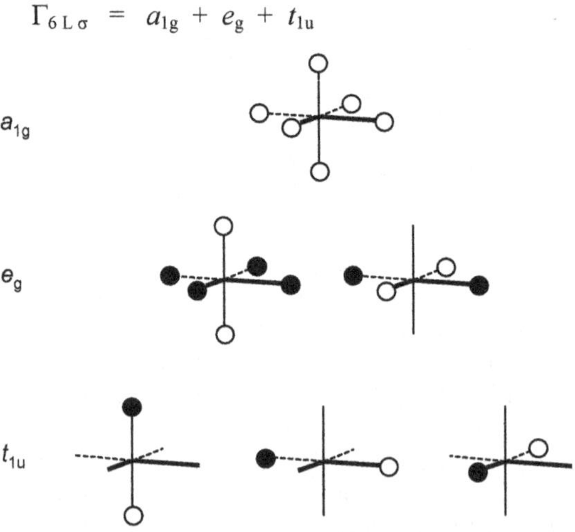

Fig. 9.2 SALCs for the σ-bonded ligands in an octahedral complex.

As we would expect, we have symmetry labels which describe the six possible combinations of six ligand orbitals. The labels tell us that these six SALCs are grouped into a degenerate set of three (t_{1u}), a second degenerate set of two (e_g) and a unique combination (a_{1g}). These six SALCs are shown in Fig. 9.2. Exact mathematical descriptions of the SALCs can be obtained using a mathematical technique known as **projection operators**, and a basic introduction to this technique is given in Appendix 1. However, the origin of the SALCs can be broadly appreciated in a pictorial manner by looking at combinations of the two SALCs which describe the linearly-arranged pair of ligands (L_2) with the cyclic arrangement of four co-planar ligands (L_4), both of which have been described in earlier chapters.

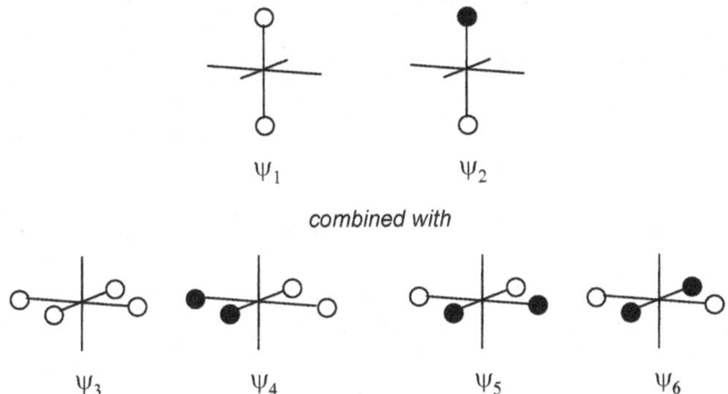

Fig. 9.3 Ligand SALCs for an octahedral complex generated by combining SALCs for L_2 and L_4.

Firstly, ψ_2 has a horizontal nodal plane and thus has no interaction with any of the SALCs ψ_3 to ψ_6; there will always be equal amounts of bonding and anti-bonding above and below the horizontal plane. ψ_1, however, can combine in-phase and out-of-phase with ψ_3, both of which have no nodes, to generate the a_{1g} ($\psi_1 + \psi_3$) and one of the e_g pair of L_6 SALCS ($\psi_1 - \psi_3$). ψ_6, which has two nodes and therefore no symmetry match with either of the SALCs ψ_1, ψ_2 (which have only one node), makes up the other component of the e_g pair.

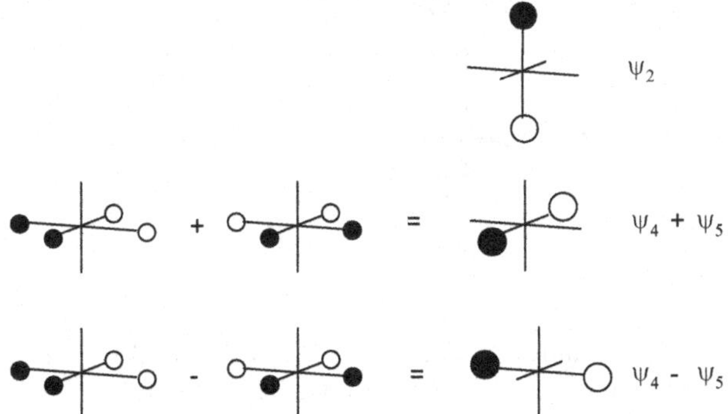

Fig. 9.4 Origin of the SALCs of t_{2g} symmetry for a set of six octahedral ligands.

Although it is not obvious from Fig. 9.3, the SALCs ψ_2, ψ_4 and ψ_5 are all equivalent and make up the t_{2g} set of L_6 SALCs (Fig. 9.4). The latter two look somewhat different, but a symmetry-equivalent degenerate pair such as ψ_4 and ψ_5 can be replaced by an equally valid "sum and difference" pair of orbitals.

9.2 *d*-ORBITAL SYMMETRY LABELS

Before we can consider the bonding in octahedral transition metal complexes we need to know more about the symmetry labels for the five *d*-orbitals under O_h symmetry. From other courses in inorganic chemistry you are probably aware that in an octahedral field (i.e. an octahedral set of ligands) the *d*-orbitals can be grouped into a degenerate set of three (t_{2g}) and a degenerate set of two (e_g). You can confirm this for yourself in the following exercise:

SAQ 9.1 : Complete the following reducible representation and hence show that the five d-orbitals are labelled t_{2g} and e_g under O_h symmetry. Count 1, 0, -1 if the d-orbital is unmoved, moved or reversed under a given operation.

O_h	E	$8C_3$	$6C_2$	$6C_4$	$3C_2{}^a$	i	$6S_4$	$8S_6$	$3\sigma_h$	$6\sigma_d$
$\Gamma_{d\text{-orbitals}}$		-1					-1	-1		

[a] $C_2 \equiv C_4^2$.

Answers to all SAQs are given in Appendix 3.

Consulting the O_h character table (Appendix 5) reveals that listed in the same row as the T_{2g} label are the binary combinations *xy*, *xz*, *yz* and in the same row as E_g the binaries $2z^2\text{-}x^2\text{-}y^2$, $x^2\text{-}y^2$. Firstly, this confirms that the d_{xy}, d_{xz} and d_{yz} form the t_{2g} set and that d_{z^2} and $d_{x^2\text{-}y^2}$ make up the e_g pair; secondly, it reveals the following shortcut in assigning symmetry labels:

- *d*-orbitals have the same symmetry as that of the corresponding binary function and can be read directly from the character table.

SAQ 9.2 : Octahedral transition metal complexes can undergo tetragonal distortions to lead ultimately to a square-planar complex. How do the d-orbital symmetries change under this distortion ?

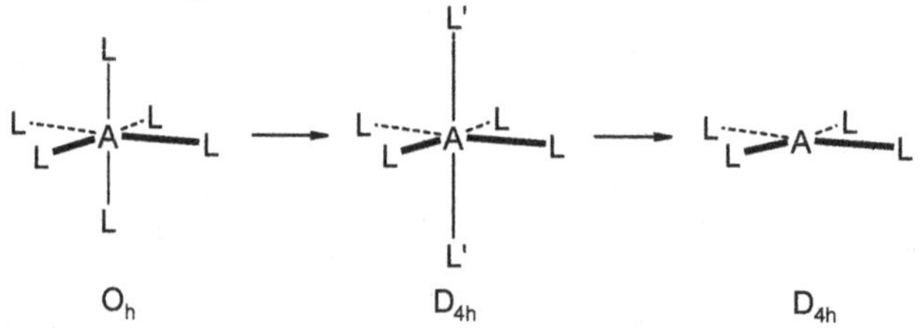

We can now summarise the findings of Sections 9.1 and 9.2 by drawing all the matches between ligand SALCs and metal AOs for an octahedral complex (Fig. 9.5).

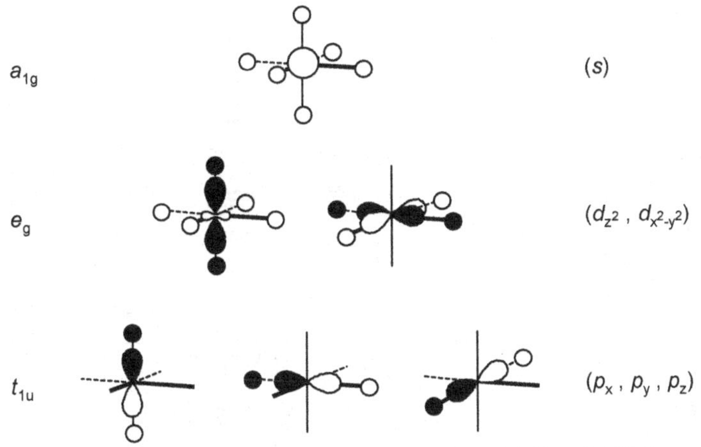

a_{1g} (s)

e_g $(d_{z^2}, d_{x^2-y^2})$

t_{1u} (p_x, p_y, p_z)

Fig. 9.5 *d*-orbitals and their symmetry matches with ligand SALCs.

9.3 OCTAHEDRAL P-BLOCK COMPLEXES

It is well known that the chemistry of the second row elements (Li – Ne) differs from that of the subsequent rows of the P-Block. One of the manifestations of this is that second row elements *rarely* exhibit coordination numbers in excess of four, while for elements in the subsequent rows coordination numbers in excess of this are commonplace and can be as high as 12 for certain elements. One of the older explanations for this is that each bond is described by an electron pair and thus the central atom requires more than four available orbitals for bonding if coordination numbers greater than four are to be achieved. For the second row elements, only the 2*s* and the three 2*p* orbitals are available, so coordination numbers maximise at four. From row three onwards (Na), the *d*-orbitals become available, so higher coordination numbers are possible. However, second row elements do, on occasion, exceed four-coordination, and, in addition, the energy of the *d*-orbitals is significantly higher than that of the *s*- and *p*-orbitals of the same principal quantum number.

The whole question of *d*-orbital participation in the chemistry of main group element compounds is open to debate, so this section is devoted to a bonding description of octahedral main group element compounds, with and without *d*-orbital participation. This analysis also impinges on the concept of **hypervalency**, when a main group element forms compounds in which the normal octet of valence electrons is exceeded; species such as PF_5 and SF_6 fall into this category.

We already know the symmetry of the six ligand SALCs under O_h symmetry (a_{1g}, e_g, t_{1u}) as well as the symmetry labels for the *d*-orbitals on the central atom (t_{2g}, e_g). To complete all the information that is needed, the symmetry labels for the *s*- and *p*-orbitals on the central atom can be read directly from the O_h character table:

$$s \ : \ a_{1g} \qquad\qquad p_x, p_y, p_z \ : \ t_{1u}$$

Considering a third row complex such as, for example, $[AlCl_6]^{3-}$ or SF_6, the 3*s* and 3*p* orbitals lie at lower energy than the 3*d*. Pairing up ligand SALCs and central atom AOs of the same symmetry generates the MO diagrams shown in Fig. 9.6.

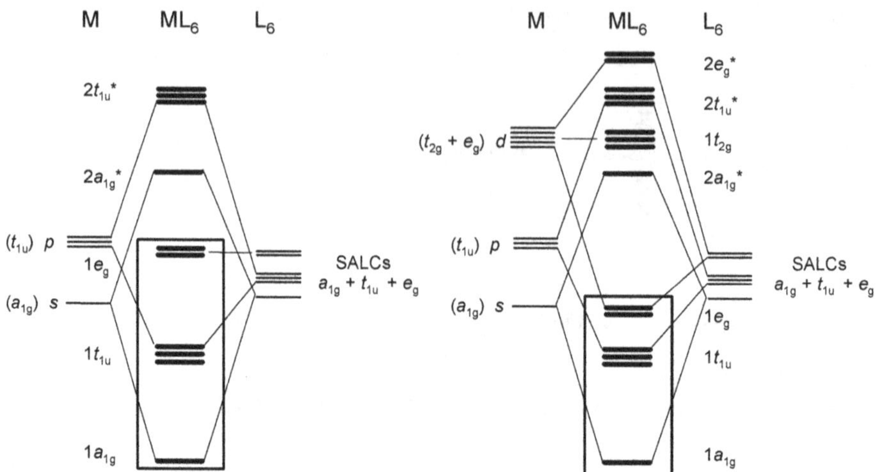

Fig. 9.6 MO diagrams for a σ-bonded octahedral complex of a main group element, (a) without *d*-orbital involvement and (b) including *d*-orbitals.

Without the inclusion of *d*-orbitals, the 12 valence electrons (six electron pairs) occupy the bonding a_{1g}, t_{1u} and non-bonding e_g MOs; there is no need to involve *d*-orbitals to rationalise the bonding (Fig. 9.6a). However, if *d*-orbitals are included, the e_g pair (d_{z^2} and $d_{x^2-y^2}$) can combine with the e_g pair of SALCs to produce bonding and anti-bonding pairs of MOs (Fig. 9.6b). The key result of this is that the non-bonding e_g pair of MOs take on bonding character. Thus, while *d*-orbitals are not required to explain coordination numbers higher than four for main group element compounds (hence the observation that lithium, for example, can exhibit coordination numbers as high as seven), their participation, should the *p-d* energy gap be sufficiently small, increases the overall stability of the species.

9.4 OCTAHEDRAL TRANSITION METAL COMPLEXES

If we consider only ligands which are σ-bonded to the central transition metal, the MO diagram for an octahedral transition metal complex is closely related to that of Fig. 9.6b, the only modification being that the *d*-orbitals are now lower in energy than the *s*- and *p*-orbitals (Fig. 9.7). The twelve electrons associated with the ligands (each as a two-electron donor) just fill the $1a_{1g}$, $1t_{1u}$ and $1e_g$ MOs, just as with their main group element analogues. The $1t_{2g}$ and $2e_g$* are occupied by the *d*-electrons of the M^{n+} moiety and are separated by an energy Δ_o, which is the octahedral crystal field splitting energy in CFT parlance. In fact, this latter part of the MO diagram is qualitatively the same picture as that generated by CFT, with the exception that in an MO approach e_g is an anti-bonding pair of orbitals while no such distinction is made in the electrostatic-based CFT analysis.

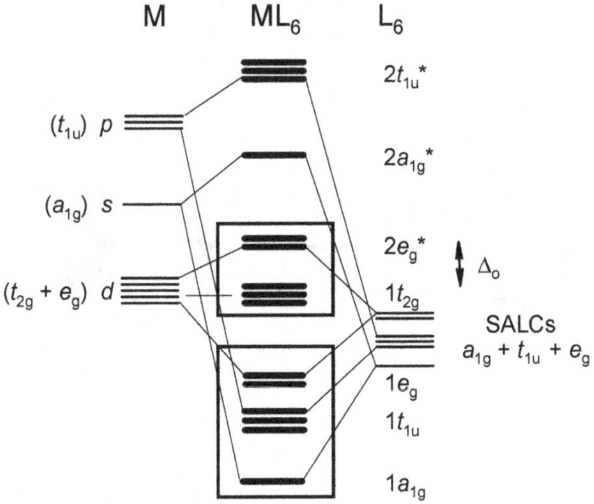

Fig. 9.7 MO diagram for an octahedral complex of a transition element, assuming only σ-bonded ligands.

> *SAQ 9.3 : Derive an MO diagram for a tetrahedral transition metal complex ML_4, considering only σ-bonding. Correlate the diagram with CFT.*

9.5 π-BONDING AND THE SPECTROCHEMICAL SERIES

The spectrochemical series is a listing of ligands based on the magnitude of the crystal field splitting, Δ_o (Fig. 9.7) that they induce. This list is based entirely on data derived from u.v.-visible spectra; it has no theoretical basis. A representative part of the series, including the common ligands, is:

$$I^- < Br^- < Cl^- < F^- < H_2O < NH_3 < PPh_3 < CN^- < CO$$

This list is not at all what might be expected from CFT, a theory based on electrostatic repulsions : the largest splittings are generated by neutral ligands such as CO, while small splittings are associated with highly electronegative ligands such as the halide ions. A rationale for this series can be found in ligand field theory, a more sophisticated analysis of the bonding in transition metal complexes based on an MO approach. This analysis requires us to consider π-bonding effects as well as the σ-interactions presented in Section 9.4.

Twelve *p*-orbitals, two for each ligand, are available for π-bonding with the *d*-orbitals on the transition metal (Fig. 9.8a). The symmetry labels for the twelve SALCs which describe the combinations of the ligand *p*-orbitals can be derived by group theory:

O_h	E	$8C_3$	$6C_2$	$6C_4$	$3C_2$[a]	i	$6S_4$	$8S_6$	$3\sigma_h$	$6\sigma_d$
$\Gamma_{12\,L\,\pi}$	12	0	0	0	−4	0	0	0	0	0

[a] $C_2 \equiv C_4^{\,2}$.

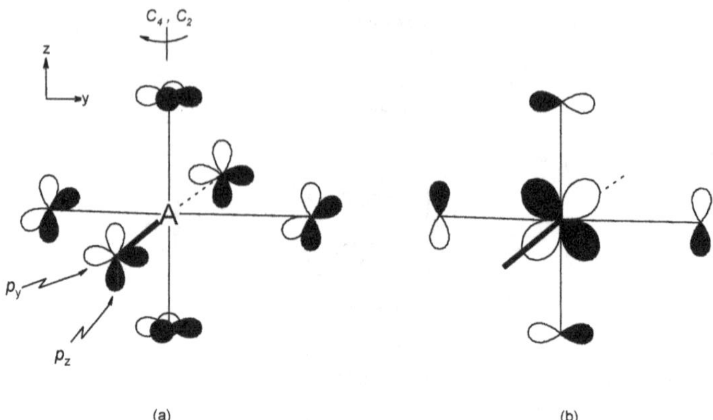

Fig. 9.8 (a) Ligand *p*-orbitals capable of π-interactions with the metal and (b) an example of one of the metal-ligand π-interactions of t_{2g} symmetry.

Under E all twelve orbitals are unmoved, while they all move under the remaining operations except C_2 (= C_4^2) and σ_h, which require further comment. For C_2, the four orbitals lying on the axis reverse (–1) while all the other orbitals move. The situation is more complex for σ_h and we will just focus on the horizontal *xy* mirror plane. The four orbitals lying along *z* all move under this σ_h and count 0. Of the eight orbitals attached to the four ligands in the *xy* plane, four lie in the mirror plane and are unmoved (1) while the other four (vertical in *z* direction) are reversed (–1) making a total contribution of 0. The reducible representation above can be converted to the following sum of irreducible representations:

$$\Gamma_{12\,L\,\pi} \;=\; t_{1g} + t_{2g} + t_{1u} + t_{2u}$$

Of these, only the t_{2g} set has any significant impact on the MO diagram. One of the three such interactions is shown in Fig. 9.8b; two other identical interactions occur in the *xz* and *xy* planes. The t_{1u} set does have a symmetry match with the metal *p*-orbitals, but the energy separation is large so the interaction will be minor and then only serve to lower the energy of the bonding t_{1u} MO. The t_{1g} and t_{2u} π-SALCs have no match and are non-bonding.

The impact of mixing the *d*-orbitals of t_{2g} symmetry (d_{xy}, d_{xz}, d_{yz}) with the π-SALCs of the same symmetry depends on the relative energies of the two sets of orbitals. If the SALCs are of lower energy than the metal AOs, which occurs when the *p*-orbitals on the ligands are filled, then the bonding t_{2g} MOs are ligand-centred and are filled; the anti-bonding t_{2g}* π-MOs are metal-centred. This effectively raises the energy of the metal t_{2g} AOs and decreases the magnitude of Δ_o (Fig. 9.9a). Ligands which behave in this manner are call **π-donors**, as they donate a pair of electrons in forming the π-bond. They are typified by the halide ions, whose lone pairs are in *p*-orbitals which are suitably oriented.

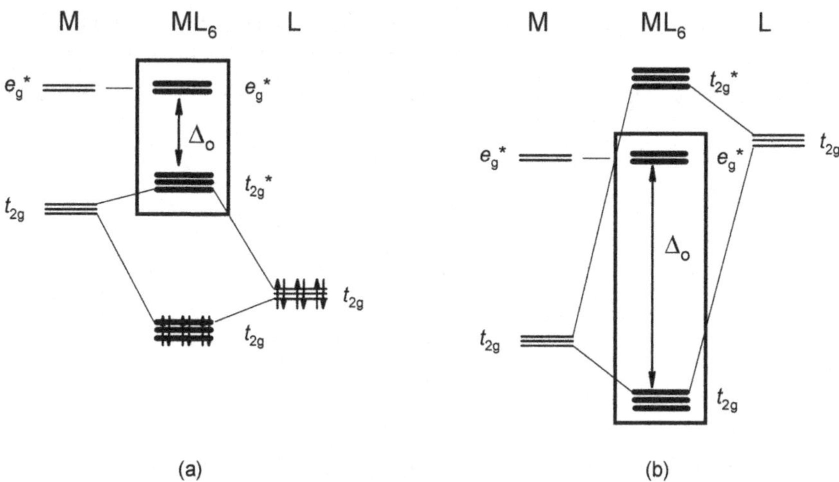

Fig. 9.9 The effect of π-bonding on Δ_o (a) when the ligands are π-donors and (b) when the ligands are π-acceptors. Boxes show the orbitals which are occupied by the d^n electrons.

Conversely, when the ligand has an empty orbital available to receive electrons from the metal in π-bond formation it is termed a **π-acceptor**. The ubiquitous example of a π-acceptor ligand is CO, where the empty orbital is the π* associated with the C-O bond. As an anti-bonding orbital, it is at higher energy than the metal t_{2g} set. The bonding t_{2g} π-MO that is formed when SALCs and AO mix is largely metal-centred, while t_{2g}* is closely related to the ligand π*. This interaction has the effect that Δ_o increases (Fig. 9.9b). The spectrochemical series, far from being dependent on electrostatic interactions, is a reflection of the π-donor / π-acceptor character of the ligands.

9.6 SUMMARY

- the six ligand SALCs which describe an octahedral arrangement of ligands have the symmetry $a_{1g} + e_g + t_{1u}$.
- *d*-orbitals have the same symmetry as the corresponding binary function and can be read directly from the character table.
- main group element compounds can form octahedral complexes without the involvement of *d*-orbitals on the central atom; the e_g ligand SALCs are non-bonding.
- if the energy of the *d*-orbitals is sufficiently low to involve them in the bonding of an octahedral main group element compound, they enhance the bonding by making the non-bonding e_g ligand SALCs bonding in character (match with d_{z^2}, $d_{x^2-y^2}$ of e_g symmetry).

- for transition metal complexes, the crystal field spitting (Δ_o) is the separation between t_{2g} and e_g* MOs.
- π-bonding between ligands and a transition metal involves orbitals of t_{2g} symmetry on the two centres and decreases Δ_o if the ligand is a π-donor.
- π-bonding between ligands and a transition metal increases Δ_o if the ligand is a π-acceptor.

PROBLEMS

Answers to all problems marked with * are given in Appendix 4.

1*. Sketch the orbital interactions which describe the bonding between the *d*-orbitals on iron and the π-bonds in benzene in the complex $(C_6H_6)Fe(CO)_2$; assume local C_{6v} symmetry for the $(C_6H_6)Fe$ part of the molecule.

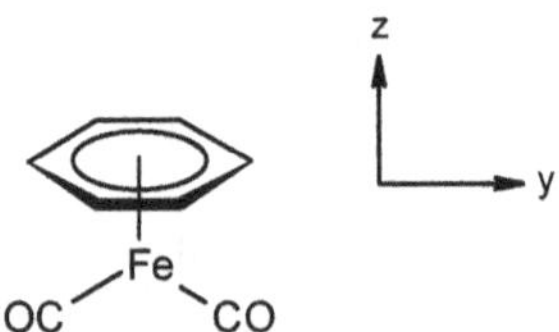

To do this you should:

- identify the symmetry labels for the $\pi(p_z)$ SALCs on benzene.
- assign these SALCs to the six nodal arrays for cyclic H_6 (*see* Fig. 8.8 and problem 6, Chapter 8).
- identify the symmetry labels for the 3*d* orbitals on iron.
- match SALCs and metal *d*-orbitals of the same symmetry.

2. Derive an MO diagram for a square planar (D_{4h}) transition metal complex ML_4, assuming only σ-bonding (see Fig 5.1 for location of symmetry elements):

- identify the symmetry labels for the four ligand SALCs.
- identify the symmetry labels for the *s*-, *p*- and *d*-orbitals.
- generate an MO diagram, taking into account the relative energies of the AOs.

3. What is the difference between the relative energies of the MOs which formally contain the metal *d*-electrons in the MO diagram for a square-planar (D_{4h}) transition metal complex (Question 2) and that derived by CFT ?

   Refine the MO diagram of Question 2 by considering the π-interactions between the four ligands and the metal. You will need to identify the symmetry labels for the SALCs associated with the eight *p*-orbitals oriented for π-bond formation with the metal.

4. Cyclobutadiene can be stabilised by bonding to a transition metal, such as in the compound $(C_4H_4)Fe(CO)_3$. Assuming local C_{4v} symmetry for the $(C_4H_4)Fe$ part of the molecule:

   - derive the symmetry labels for the four SALCs on C_4H_4 (p_z).
   - identify the symmetry labels for the AOs on iron under C_{4v} symmetry.
   - match AOs and SALCs of the same symmetry and confirm the matches with sketches of the relevant combinations.

5*. What are the symmetry labels for the fluorine SALCs which are available for σ-bonding in XeF_4 (D_{4h} ; *see* Fig. 5.1 for the location of the symmetry elements) ?

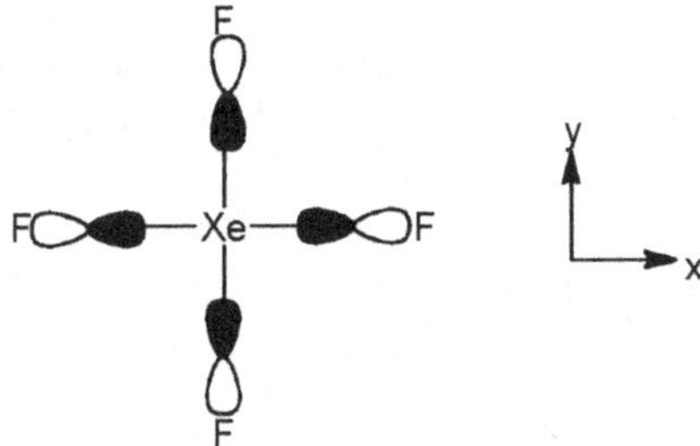

   What are the symmetry labels for the valence *s*-, *p*- and *d*-orbitals on xenon ?

   Adapt the nodal patterns for cyclic H_4 to draw the four fluorine SALCs and, by matching symmetry labels beween SALCs and AOs, draw the four σ-bonding MOs for XeF_4.

6. The σ-bonding in hypervalent five-coordinate complexes AL_5, e.g. PCl_5, can be analysed in the same way as was shown for octahedral complexes, without the need for *d*-orbital involvement. Furthermore, in trigonal bipyramidal complexes of this type it is well-established that the two axial bonds are slightly longer than the three equatorial ones.

   Rationalise these observations by completing the MO diagram given overleaf, adding symmetry labels to the AOs on A and the five ligand SALCs, then matching orbitals of the same symmetry to generate MOs.

You should:

- determine the symmetry labels for the *s*- and *p*-AOs on the central atom (A).
- determine the symmetry labels for the five ligand SALCs.
- draw these SALCs, by combining, where possible, the axial and equatorial SALCs shown below:

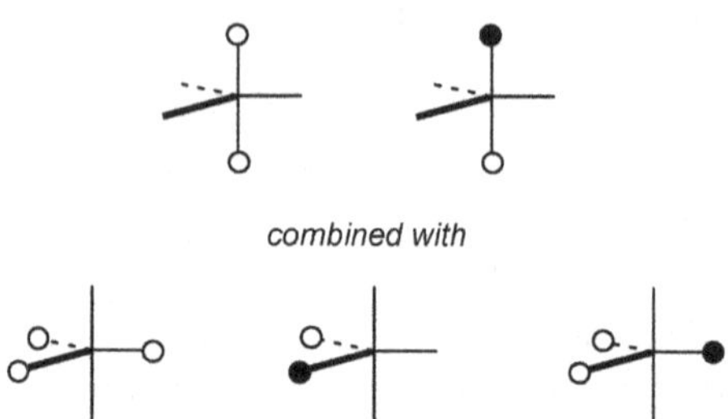

(*Hint: the out-of-phase combination of axial orbitals has a nodal plane which coincides with the equatorial SALCs; SALCs with a or b labels can be identified in the usual way by counting 1, 0, -1 for the effect of each symmetry operation*)

- label the MO diagram (*below*) and add the ten electrons associated with the five σ-bonds.
- calculate the average bond order for the axial and equatorial bonds and hence rationalise why the axial bonds are slightly longer than the equatorial ones. (*Hint: for each MO, is it associated with only the axial bonds, only the equatorial bonds, or both ?*)

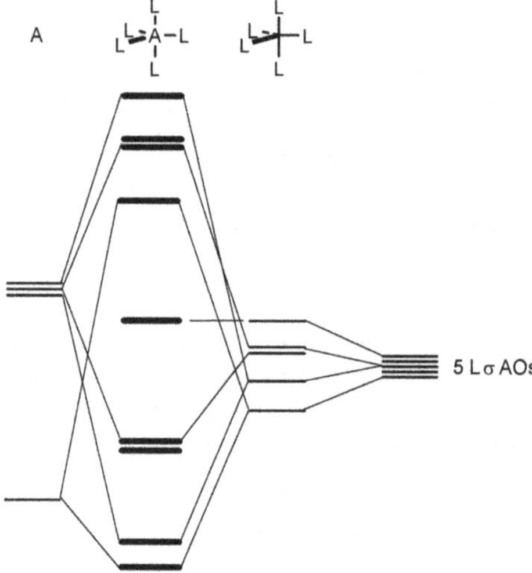

10
FERROCENE

As with Part II, Part III of the book concludes with a complete worked example which brings together all the concepts and techniques described in earlier sections. The molecule which forms the focus of this chapter is ferrocene, $(C_5H_5)_2Fe$, the parent in a family of so called "sandwich" compounds and perhaps the most studied and exploited organometallic compound known (Fig. 10.1).

Part III began with a look at the simple problem of constructing MO diagrams for H_2 and H_3, which we carried out in a purely descriptive manner. Armed with the techniques of group theory outlined in previous chapters, we conclude this part of the book with a far more complex problem : describing the bonding in ferrocene by identifying orbital interactions between the ten *p*-orbitals on carbon and the nine AOs (4*s*, 4*p*, 3*d*) on iron.

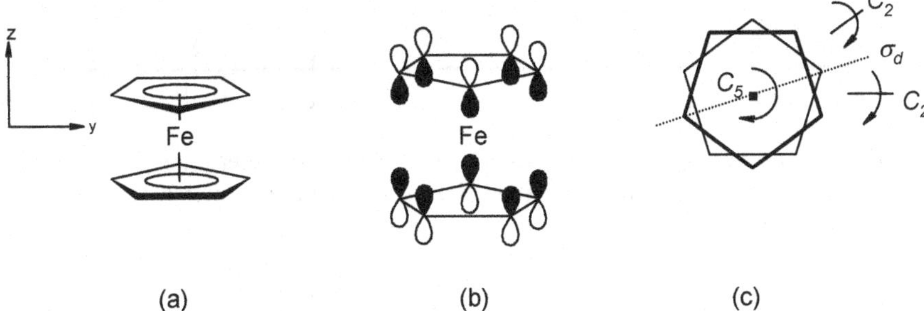

Fig. 10.1 Ferrocene, (a) showing delocalisation within the cyclopentadienyl rings, (b) the p_z orbitals on carbon used in bonding and (c) the location of key symmetry elements in D_{5d}.

The symmetry of ferrocene is D_{5d}, in which the two, planar, cyclopentadienyl rings are staggered with respect to each other.* Each ring has a delocalised π-system

* The staggered (D_{5d}) and eclipsed (D_{5h}) geometries are close in energy (10–20 $kJmol^{-1}$) and both conformations occur randomly in the crystalline form.

built from the p_z orbitals on carbon. Both rings bond symmetrically to the iron through each of the five carbon atoms, in what is termed an η^5- ("eta-5") mode.

10.1 CENTRAL ATOM ORBITAL SYMMETRIES

The symmetries of the 3*d*-, 4*s*- and 4*p*-orbitals on iron can be read directly from the D_{5d} character table, remembering that *s*-orbitals are always perfectly symmetrical, *p*-orbitals have the symmetry of the corresponding $\mathbf{T_x}$, $\mathbf{T_y}$, $\mathbf{T_z}$ and *d*-orbitals have the symmetry of their binary function:

$4s$	: a_{1g}	d_{z^2}	: a_{1g}
$4p_x$, $4p_x$	: e_{1u}	$d_{x^2-y^2}$, d_{xy}	: e_{2g}
$4p_z$	: a_{2u}	d_{xz}, d_{yz}	: e_{1g}

10.2 SALCS FOR CYCLOPENTADIENYL ANION

The key symmetry elements for D_{5d} are shown in Fig. 10.1c. The principal axis (C_5) passes through the iron perpendicular to the C_5H_5 rings; the S_{10} axis is coincident with C_5. There are five C_2 axes perpendicular to C_5 and which also pass through the metal centre, along with five vertical planes (σ_d).

In the character table, the operations listed as $2C_5$ and $2C_5^2$ refer to $C_5^{1,4}$ and $C_5^{2,3}$, respectively. Many of the operations associated with the S_{10} axis are equivalent to other operations and appear elsewhere in the table ($S_{10}^2 \equiv C_5^1$, $S_{10}^4 \equiv C_5^2$, $S_{10}^6 \equiv C_5^3$, $S_{10}^8 \equiv C_5^4$, $S_{10}^{10} \equiv E$, $S_{10}^5 \equiv i$), leaving $2S_{10}^3$, which corresponds to $S_{10}^{3,7}$, and $2S_{10}$, which relate to $S_{10}^{1,9}$. Luckily, these subtle distinctions don't affect the determination of the reducible representation.

The reducible representation for the ten p_z orbitals under D_{5d} symmetry is found by counting 1, 0, -1 for unmoved, moved and reversed orbitals, in the usual way:

D_{5d}	E	$2C_5$	$2C_5^2$	$5C_2$	i	$2S_{10}^3$	$2S_{10}$	$5\sigma_d$
$\Gamma_{10\,C\,pz}$	10	0	0	0	0	0	0	2

The only operation which leaves any orbital unmoved, other than E, is σ_d, which contains one p_z from opposite sides of each ring. Using the reduction formula (Section 3.2) :

$$\Gamma_{10\,C\,pz} = a_{1g} + e_{1g} + e_{2g} + a_{2u} + e_{1u} + e_{2u}$$

As a check, we expect ten SALCs from the combination of ten AOs, and we have generated the required number of symmetry labels, remembering that *e* is a degenerate pair.

It should be noted at this stage that, from a comparison of the labels for the ten p_z SALCs with those for the AOs on iron, that two of the former, those of e_{2u} symmetry, have no matching AO on the metal.

The ten C(p_z) SALCs can be pictured by combining each of the five possible SALCs for a single cyclic H_5 system (which in turn are generated from the corresponding linear H_5 arrays) with itself, in both in-phase and out-of-phase fashions. With reference to Fig. 8.7, the five cyclic nodal arrays for five p_z orbitals are:

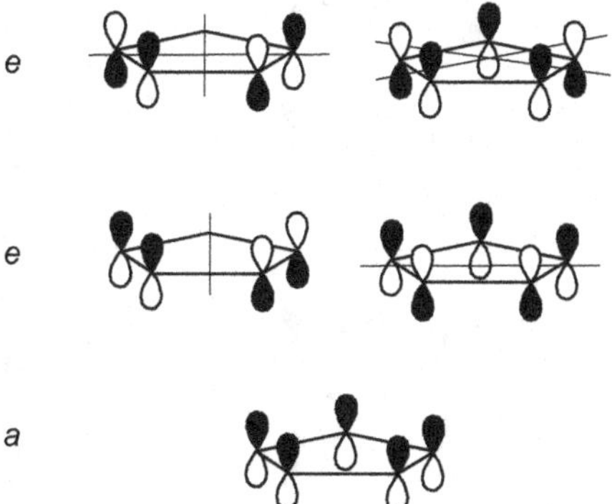

Fig. 10.2 SALCs for cyclic arrangements of five p_z orbitals.

Without determining the symmetry labels of the SALCs in Fig. 10.2 explicitly, we can guess that the perfectly symmetrical SALC will have *a* symmetry, while the pairs of SALCs with the same number of nodes will group into two *e* pairs. This is relatively easy to confirm by considering the SALCs for a single C_5H_5 ring, which has D_{5h} symmetry:

D_{5h}	E	$2C_5$	$2C_5^2$	$5C_2$	σ_h	$2S_5$	$2S_5^3$	$5\sigma_v$
$\Gamma_{5\,C\,pz}$	5	0	0	–1	–5	0	0	1

$$\Gamma_{5\,C\,pz} = a_2'' + e_1'' + e_2''$$

The symmetry labels are now more complete as they relate to a specific point group, but they serve to confirm the basic groupings and symmetry labels of the SALCs shown in Fig. 10.2.

In order to produce the ten SALCs associated with the pair of parallel cyclopentadienyl rings in ferrocene, we combine pairs of identical SALCs for the individual rings, in both in-phase and out-of-phase combinations (Fig. 10.3).*

The symmetry labels which describe the representation $\Gamma_{10\,C\,pz}$ (a_{1g}, e_{1g}, e_{2g}, a_{2u}, e_{1u}, e_{2u}) can only easily be partially assigned to the SALCs in Fig. 10.3 at this stage. That is, if a SALC is generated by combination of two *a* symmetry C_5H_5 SALCs then it will retain an *a* label; similarly for combinations of C_5H_5 SALCs which have *e* symmetry. In addition, the subscripts *g* and *u* (symmetric and anti-symmetric with respect to inversion, respectively) allow these designations to be added. As a consequence, both the a_{1g} and a_{2u} SALCs can be fully labelled, though only partial labels can be assigned to the eight SALCs of *e* symmetry.

---

* Combining pairs of SALCs in this way requires that the nodal planes are coincident (*see Section 7.3*), so ruling out combinations of SALCs with orthogonal nodal planes. Thus, the *e* pair of SALCs only combine "like-with-like" and not with each other.

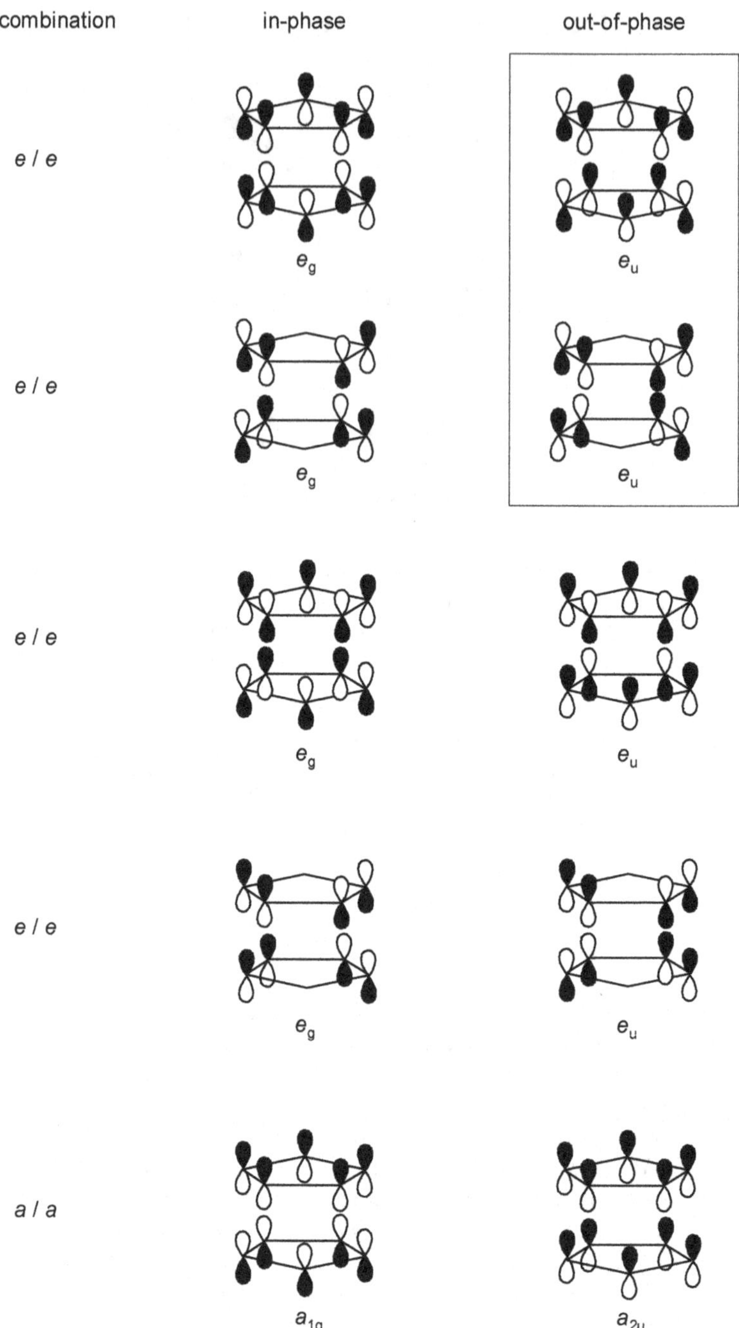

Fig. 10.3 The ten SALCs for pairs of C_5H_5 rings, generated by combining pairs of identical SALCs for an individual C_5H_5 ring, both in-phase and out-of-phase; only partial symmetry labels have been added for the *e* SALCs. SALCs with no matching AO on iron are boxed.

10.3 MOLECULAR ORBITALS FOR FERROCENE

All that remains is to combine the ligand SALCs with the AOs on iron of identical symmetry; the types of ligand-iron bond that can be formed by the different orientations of orbital interaction have been reviewed in Section 6.2.

In the following two cases, the matches will be unambiguous, as the a_{1g} and a_{2u} SALCs have been identified:

Table 10.1 Symmetry-allowed orbital combinations between cyclopentadiene SALCs and iron AOs for ferrocene (D_{5d}).[a]

SALC	AO	Label	MOs	M-L Bond type
	$4s$, $3d_{z^2}$	a_{1g}	 	σ
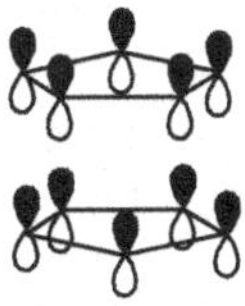	$4p_z$	a_{2u}	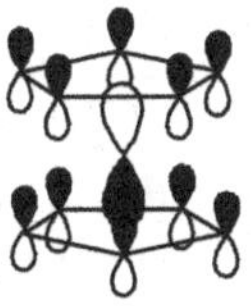	σ

[a] Only the bonding combination for each SALC and AO pair is shown.

In the remaining cases it is a question of visually making the appropriate matches, starting from the AOs on iron of known symmetry and pairing them with the appropriate SALC generated from the two cyclopentadienyl rings:

Table 10.1 ***continued***

SALC	AO	Label	MOs	M-L Bond type
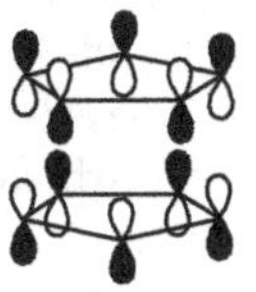	$3d_{xz}$	e_{1g}		π
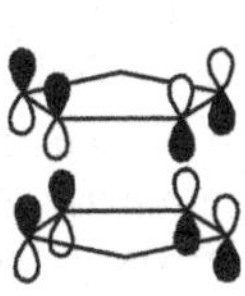	$3d_{yz}$	e_{1g}	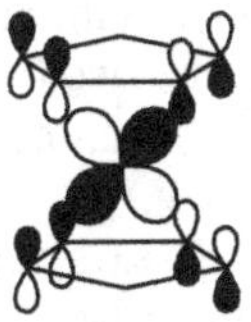	π
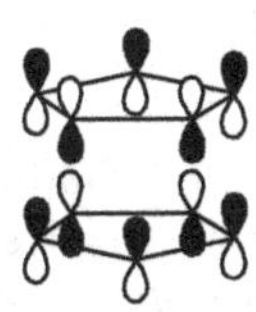	$4p_x$	e_{1u}	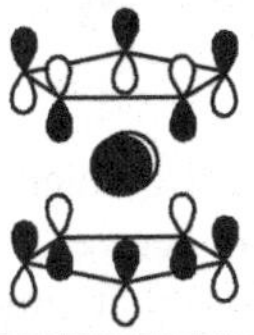	π

SALC	AO	Label	MOs	M-L Bond type
	$4p_y$	e_{1u}	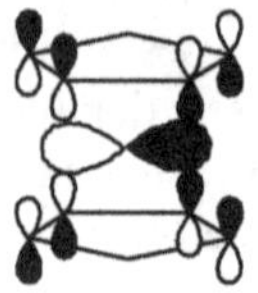	π
	d_{xy}	e_{2g}	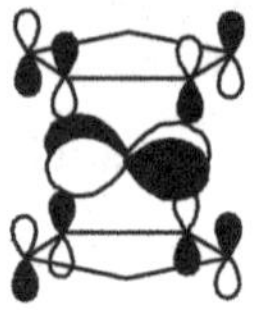	δ
	$d_{x^2\text{-}y^2}$	e_{2g}	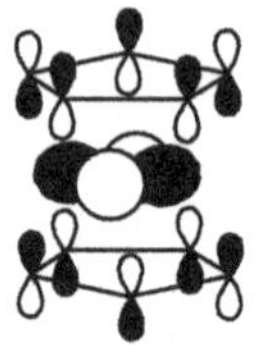	δ

[a] Only the bonding combination of SALC and AO is shown.

Note how the vertical nodes in the ring SALCs (the horizontal node across the C_5 ring is not relevant here) coincide with those of the matching AO.

By elimination, the SALCs highlighted in the box of Fig. 10.3 make up the e_{2u} pair and have no symmetry match with a valence AO on iron.

At this point, group theory has completed its contribution to building up a bonding scheme for ferrocene. It has guided us through the complex way in which ten p_z AOs on carbon and nine AOs on iron combine together to give a set of MOs which describe the bonding in this most important of sandwich compounds. To generate an MO diagram requires us to know the energies of the ten p_z SALCs and the 3*d*-, 4*s*- and 4*p*-orbitals on iron, and the extent to which symmetry-matched orbitals interact, as this determines the extent of stabilisation / destabilisation of the bonding / anti-bonding MOs produced. In turn, this demands a detailed computational study which is beyond the scope of our qualitative approach. Indeed, the sequence of MOs has been the subject of several such studies but still gives rise to controversy. For completeness, however, a qualitative MO diagram, which represents the commonly accepted order of energy levels, is shown in Fig. 10.4.

Firstly, on the left hand side of the diagram are the symmetry labels for the ten SALCs which arise from combining, under the D_{5d} symmetry of the molecule, the ten p_z AOs on carbon. Since the two C_5 rings are *ca.* 4 Å apart, there will be a negligible interaction between them, meaning that in-phase and out-of-phase combinations of identical C_5 SALCs e.g. a_{1g}/a_{2u}, e_{1g}/e_{1u} or e_{2g}/e_{2u}, have essentially identical energies. The energies of the ten SALCs for the C_{10} unit increase as the number of nodes increases, so they lie in the order:

$$a_{1g} \approx a_{2u} < e_{1g} \approx e_{1u} < e_{2g} \approx e_{2u}$$

On the right hand side of the diagram are the AOs on iron, in order $3d < 4s < 4p$ of increasing energy. In the centre of the diagram is a schematic showing the relative energies of the MOs, without attempting to accurately reflect energy differences. As well as showing the symmetry label for each MO, the latter is further labelled according to the type of M-L bond that is generated, allowing an additional method of identification.

The a_{1g} SALC, $3d_{z^2}$ and 4*s* orbitals all combine to give rise to three σ-MOs of a_{1g} symmetry, one of which is bonding, one non-bonding and one anti-bonding in character, and these are labelled 1σ, 3σ and 4σ* in Fig. 10.4. Since there is a sizeable energy difference between the a_{1g} SALC and the 3*d* / 4*s* AOs, the bonding 1σ MO is largely centred on the cyclopentadienyl rings. The other low energy MO is the σ-bonding combination of a_{2u} symmetry involving p_z on iron (2σ in Fig. 10.4). By the same reasoning as above, 2σ is also essentially ligand-centred. The next lowest energy MOs will be first the degenerate pair of π-MOs of e_{1g} symmetry (1π), followed by an analogous e_{1u} pair (2π); their anti-bonding partners appear as 3π* and 5π*. These π-bonding MOs have significant *d*-orbital contributions as the energy differences between the AOs and the e_{1g}/e_{1u} SALCs are less than for the contributors to the σ-MOs. Finally, the highest energy of the bonding orbitals are the degenerate pair of δ-MOs of e_{2g} symmetry (1δ). The two SALCs of e_{2u} symmetry, which have no symmetry match among the AOs on iron, appear as relatively high energy, completely ligand-centred MOs (4π).

The eighteen electrons (Fe : d^8; 2 x C_5H_5 = 10π electrons) just fill the nine, low-energy essentially bonding MOs, thus giving rise to a highly stable arrangement.

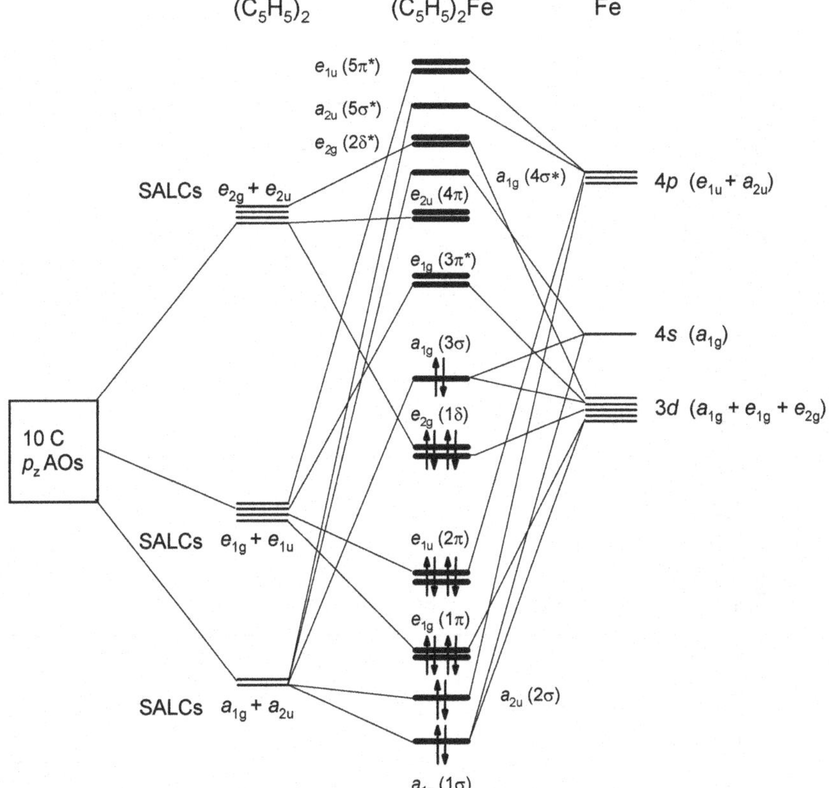

Fig. 10.4 Qualitative MO diagram for ferrocene (D_{5d}).

PROBLEMS

Answers to all problems marked with * are given in Appendix 4.

1. Sketch the orbital interactions which describe the bonding between chromium and benzene in the complex $(C_6H_6)_2Cr$ (D_{6h}). To do this you should:

   - identify the symmetry labels for the 4*s*-, 4*p*- and 3*d*-orbitals on chromium.
   - identify the symmetry labels for the 12 $\pi(p_z)$ SALCs on benzene.
   - identify which SALCs are not used in bonding to chromium.
   - match SALCs and AOs of the same symmetry and complete the following Table (following the style of Table 10.1), sketching only the bonding MO combinations:

SALC	AO	Label	MO	M-L Bond type
 z, y				
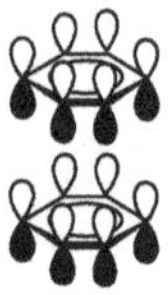				
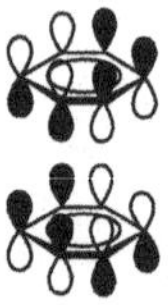				
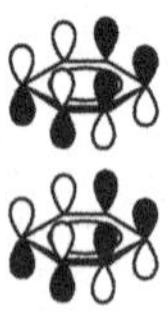				
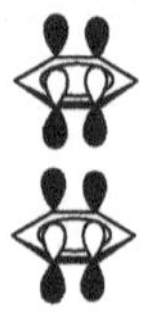				

Question 1, continued.

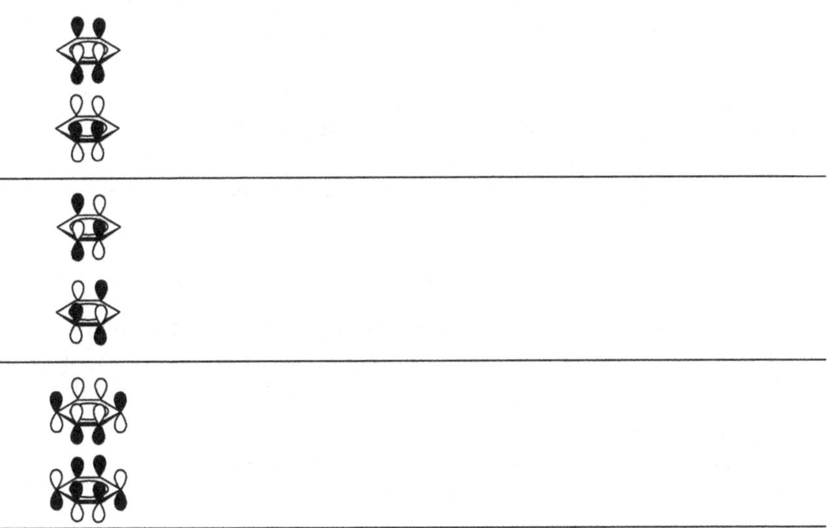

2*. The bonding in the bis-allyl complex η^3-$(C_3H_5)_2Ni$ (C_{2h}) is complex due to a large degree of mixing between AOs and SALCs. In this problem, you are only asked to identify the *possible* symmetry-allowed matches between the π-orbitals on the allyl groups and the AOs on nickel:

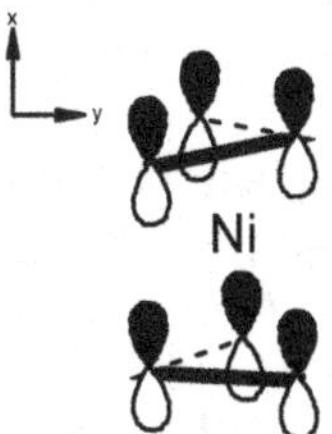

- determine the symmetry labels for the six SALCs associated with the p_z AOs on the carbons of the two allyl groups.
- draw each of the six SALCs, by combining pairs of identical SALCs for each C_3H_5 ligand (see *SAQ 6.2*) in-phase and out-of-phase.
- assign symmetry labels to each SALC.
  (*Hint : since the SALCs have a or b labels, this can be done by treating each SALC as a complete unit, counting 1, 0, -1 if it is unmoved, moves or is reversed for each operation*).
- identify the symmetry labels for the 4*s*-, 4*p*- and 3*d*-AOs on nickel.
- match the AOs on nickel with the appropriate SALCs of the two allyl groups, looking for the best match from those available.

3. $[Te_6]^{4+}$ adopts a trigonal prismatic structure (D_{3h}) similar to that of prismane, C_6H_6, an isomer of benzene.

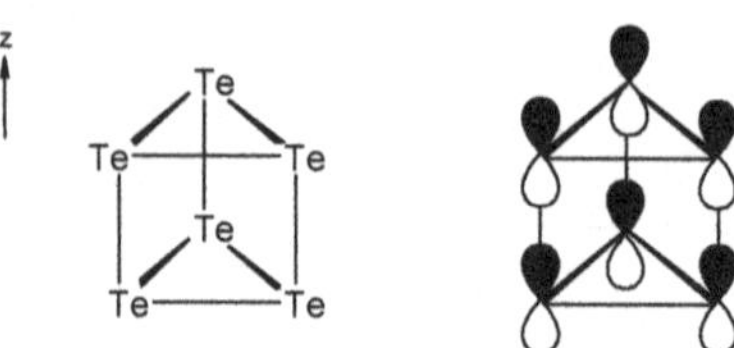

The Te-Te bonds in the triangular face are, however, shorter than those in the square face. Rationalise this observation by constructing a partial MO diagram for $[Te_6]^{4+}$ which describes the MOs involved in the σ-bonds between triangular faces.

- determine the symmetry labels for the π-SALCs associated with a Te_3 triangle (D_{3h}).

- determine the symmetry labels for the MOs which arise from combining the six p_z AOs of a Te_6 prismane.
- draw these six MOs by combining identical π-SALCs from two Te_3 units.
- draw a partial MO diagram for $[Te_6]^{4+}$ describing the bonds between the triangular faces, labelling the MOs.
  (*Hint on electron counting: within the triangular face, each Te forms two conventional 2c-2e Te-Te bonds and has a lone electron pair; remaining electrons are used to join the triangular faces*).
- what is the bond order for the bonds linking triangular faces of $[Te_6]^{4+}$?

In contrast, in prismane all the C-C bonds are equal. Consider the number of electrons now used to join the two triangular faces and hence adapt the MO diagram for $[Te_6]^{4+}$ to explain this observation.

PART IV

APPLICATION OF GROUP THEORY TO ELECTRONIC SPECTROSCOPY

11

SYMMETRY AND SELECTION RULES

In this final Part of the book, the application of symmetry to understanding and interpreting electronic spectra will be addressed. In electronic spectroscopy electrons are moved between molecular orbitals in a manner dictated by symmetry, under the influence of ultra-violet or visible radiation. In this the first of three chapters on the topic, the way the distribution of electrons in orbitals can be described in terms of symmetry labels will be addressed and we will take a preliminary look at the selection rules for electronic transitions; in the process, the selection rules for vibrational spectra, presented only as a result but without justification in Chapter 4 (*Sections 4.2 and 4.3*), will be discussed more formally.

The final chapters of the book will build on these ideas and apply them to the complex area of the electronic spectra of transition metal complexes, which are dominated by *d–d* electron transitions.

11.1 SYMMETRY OF ELECTRONIC STATES

Any orbital, either atomic or molecular, can hold a maximum of two electrons of opposite spin, as stated in the Pauli principle. For a single electron in a non-degenerate orbital i.e. one described by an *a* or *b* label, the symmetry of the electronic configuration is the same as that of the orbital e.g. a single electron in an orbital of a_1 symmetry, written $(a_1)^1$, is designated A_1. Note that while lower case letters are used for the electronic configuration, capitals are used for their symmetry labels.

Let us now consider the MO diagram for water (C_{2v}), shown in Fig. 7.3. Ignoring the non-bonding MO of a_1 symmetry (the 2*s* orbital on oxygen) then the electronic configuration for water is $(a_1)^2(b_2)^2(b_1)^2$. How, then, do we label, in symmetry terms, the situation such as these where orbitals are occupied by two electrons of opposite spin e.g. $(a_1)^2$? In this case, the overall symmetry of the electronic state is obtained as a **direct product** of the symmetries of the individual electrons i.e. in this example $(a_1)^2 = A_1 \times A_1$.

C_{2v}	E	C_2	$\sigma(xz)$	$\sigma(yz)$	
A_1	1	1	1	1	
A_1	1	1	1	1	
$A_1 \times A_1$	1	1	1	1	$= A_1$

This rather trivial case shows what is meant by a direct product. It also exemplifies a basic principle:

- the direct product of non-degenerate systems is itself a non-degenerate system.

What is the situation for the symmetry of the two electrons in the b_2 MO ?

C_{2v}	E	C_2	$\sigma(xz)$	$\sigma(yz)$	
B_2	1	−1	−1	1	
B_2	1	−1	−1	1	
$B_2 \times B_2$	1	1	1	1	$= A_1$

This establishes the following generality :

- any filled orbital (or set of degenerate orbitals) is totally symmetric (top row of the character table).

> *SAQ 11.1 : Show that, for H_2O, the $(b_1)^2$ configuration has A_1 symmetry.*
>
> **Answers to all SAQs are given in Appendix 3.**

The overall symmetry of the $(a_1)^2(b_2)^2(b_1)^2$ ground electronic configuration for H_2O is therefore $A_1 \times A_1 \times A_1 = A_1$.

Before considering which electronic transitions are allowed by symmetry, we need to determine the symmetry labels for the plausible excited states. Considering the MO diagram for H_2O (in simplified form) we have the following two possibilities for the lowest energy transitions:

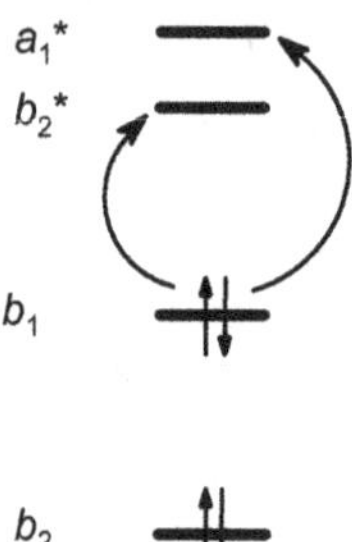

Fig. 11.1 Possible low-energy electronic transitions for H_2O (the non-bonding $1a_1$ MO shown in Fig. 7.3 omitted for clarity).

For the $b_1 \to b_2^*$ transition the excited state is $(b_1)^1(b_2^*)^1$, while for $b_1 \to a_1^*$ the excited state becomes $(b_1)^1(a_1^*)^1$. The symmetries of these excited states are determined by taking the direct product of the symmetries of the individual electrons. So, for $(b_1)^1(b_2^*)^1$, and remembering that the filled MOs (a_1, b_2) all have A_1 symmetry:

C_{2v}	E	C_2	$\sigma(xz)$	$\sigma(yz)$	
B_1	1	−1	1	−1	
B_2	1	−1	−1	1	
A_1	1	1	1	1	
$A_1 \times A_1 \times B_1 \times B_2$	1	1	−1	−1	$= A_2$

That is, the $(a_1)^2(b_2)^2(b_1)^1(b_2^*)^1$ excited state is a configuration of A_2 symmetry.

SAQ 11.2 : Show that for H_2O the $(b_1)^1(a_1^*)^1$ configuration has B_1 symmetry.

Thus, the plausible transitions are between electronic states of $A_1 \rightarrow A_2$ and $A_1 \rightarrow B_1$ symmetries. To decide if these transitions are allowed by the point group symmetry we need a more detailed look at the selection rules for such transitions.

11.2 SELECTION RULES

The electronic spectrum of water consists of a single band in the u.v. part of the electromagnetic spectrum at *ca.* 170 nm. This means that only one of the two transitions, $A_1 \rightarrow A_2$ and $A_1 \rightarrow B_2$, is allowed, but which is it ? The transitions arise from an interaction between the incident electromagnetic radiation and the molecular dipole (μ). This is also the case for vibrational spectroscopy, where the interaction between infrared radiation and the molecular dipole causes a vibrational mode to go from (usually) its ground to its first excited state. For a transition between initial (ψ_i) and final (ψ_f) energy states (either vibrational or electronic) to be **allowed**, the following integral has to be non-zero:

$$\int \psi_i \mu \psi_f \, d\tau \qquad \text{(eqn 11.1)}$$

The integral is performed over all the variables in the wavefunction (this is what is meant by "dτ"), and if the integral is zero then the transition is **forbidden**. The dipole moment is a vector quantity with components along the Cartesian axes (μ_x, μ_y, μ_z) and these have the same symmetries as the translational vectors ($\mathbf{T_x}$, $\mathbf{T_y}$, $\mathbf{T_z}$); for H_2O (C_{2v}), these correspond to B_1 ($\mathbf{T_x}$), B_2 ($\mathbf{T_y}$) and A_1 ($\mathbf{T_z}$).

Considering first the $A_1 \rightarrow B_1$ electronic transition (i.e. $\psi_i \rightarrow \psi_f \equiv b_1 \rightarrow a_1^*$ in Fig. 11.1), then the transition will be allowed if any of the following are non-zero:

$$A_1 \times \begin{pmatrix} \mu_x \\ \mu_y \\ \mu_z \end{pmatrix} \times B_1 = A_1 \times \begin{pmatrix} \mathbf{T_x} \\ \mathbf{T_y} \\ \mathbf{T_z} \end{pmatrix} \times B_1 = A_1 \times \begin{pmatrix} B_1 \\ B_2 \\ A_1 \end{pmatrix} \times B_1$$

i.e. any of the following direct products must be non-zero:

$$A_1 \times B_1 \times B_1 = A_1$$

$$A_1 \times B_2 \times B_1 = A_2$$
$$A_1 \times A_1 \times B_1 = B_1$$

The following rule is presented without rigorous proof as it is beyond the scope of this book:

- the integral is non-zero if it is unchanged over all the operations of the point group, i.e. has the symmetry of the top row of the character table.

Though not a rigorous proof, this statement can, however, be appreciated by considering as an example any non-totally symmetric row from any character table. If the characters in that row are multiplied by the number of operations in each class, then these products sum to zero. In other words, over all the operations in the point group the function maps equally onto itself and its negative, thus summing to zero. This is not true only in the case of the totally symmetric representation of each point group. For example, consider the irreducible representations A_1 and E under the point group C_{3v}:

C_{3v}	E	$2C_3$	$3\sigma_v$	
$\chi_{(IR)}E$	2	−1	0	
Number operations (n)	1	2	3	
sum($\chi_{(IR)}E \times n$)	2	−2	0	= 0
$\chi_{(IR)}A_1$	1	1	1	
Number operations (n)	1	2	3	
sum($\chi_{(IR)}A_1 \times n$)	1	2	3	= 6

Thus, a transition is allowed if at least one component of the total integral is non-zero i.e. one of the three direct products must be the totally symmetric function (top row of character table). For C_{2v} this means A_1 symmetry, so the $b_1 \rightarrow a_2^*$ transition is allowed.

SAQ 11.3 : Show that, for H_2O, the transition integral for the $b_1 \rightarrow b_2^*$ transition (i.e. $A_1 \rightarrow A_2$ symmetry) does not contain A_1 symmetry.

As the direct products associated with the $b_1 \rightarrow b_2^*$ transition are B_2, B_1 and A_2, then this transition is symmetry forbidden. The band in the u.v. spectrum of water at *ca.* 170 nm thus corresponds to the $b_1 \rightarrow a_1^*$ electronic transition, which is essentially an excitation of a non-bonding electron on oxygen into an empty anti-bonding MO of sigma symmetry.

While it is possible to evaluate the direct product of three symmetry species *ab initio* as above, the following two generalisations will simplify the task:

- the direct product of a non-degenerate irreducible representation and itself is always the totally symmetric representation.

- the direct product of two different irreducible representations is never the totally symmetric representation.

These points are also re-emphasised by the entries in Table 11.1 under the discussion in Section 11.4, below. For now, the three direct products associated with the $A_1 \rightarrow B_1$ electronic transition in water reinforce the concepts:

$$A_1 \times B_1 \times B_1 = A_1$$
$$A_1 \times B_2 \times B_1 = A_2$$
$$A_1 \times A_1 \times B_1 = B_1$$

A short-cut to deciding if a direct product contains the totally symmetric representation is to re-order the direct product $\psi_i \times \mu \times \psi_f$ as $\psi_i \times \psi_f \times \mu$ to place emphasis on evaluation of $\psi_i \times \psi_f$. Now, if the binary direct product $\psi_i \times \psi_f$ has the same symmetry as any of the components of μ, then the total ternary direct product must be totally symmetric, as the product of a non-degenerate irreducible representation and itself is always the totally symmetric representation.

11.3 THE IMPORTANCE OF SPIN

In considering the allowed electronic transitions for water, we have only focussed on the *symmetries* of the ground and excited states. When such a transition occurs, it must also be without changing the *spin* of the electron involved:

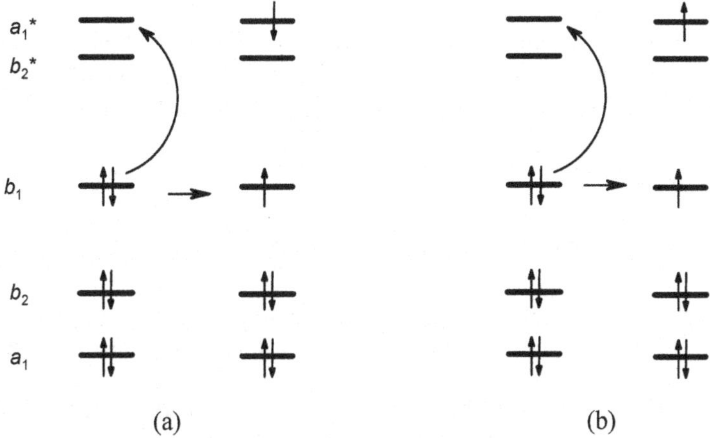

Fig. 11.2 (a) Spin-allowed (singlet → singlet) and (b) spin-forbidden (singlet → triplet) versions of the $b_1 \rightarrow a_1$* electronic transition for H_2O.

The total spin of a system (S) is the sum of all the individual electron spins (m_s); full orbitals i.e containing two electrons, count zero as the ½ and -½ spins cancel out.

$$S = \sum m_s$$

From this, the **multiplicity** of a spin state is defined as $2S+1$. Thus, a system with no unpaired electrons is a *singlet* [(2 × 0) + 1], with one unpaired electron is a *doublet* [(2 × ½) + 1], two unpaired electrons a *triplet* [(2 × 1) + 1] etc. The multiplicity of an electronic state is added as a left superscript e.g. the ground electronic state for water is $^1A_1$.

The **spin selection rule** states that:

- for a spin-allowed transition $\Delta S = 0$ i.e. there is no change in electron spin.

SAQ 11.4 : Why, in terms of spin multiplicity, is the transition shown in Fig. 11.2(a) allowed but that in (b) spin-forbidden ?

11.4 DEGENERATE SYSTEMS

With the basic principles in place, we can move to more complex situations involving transitions from degenerate MOs. Firstly, we have to look in more depth at the direct products formed when states of E or T symmetry are involved. Consider, as an example, the direct product $T_1 \times T_2$ under T_d symmetry:

T_d	E	$8C_3$	$3C_2$	$6S_4$	$6\sigma_d$	
T_1	3	0	−1	1	−1	
T_2	3	0	−1	−1	1	
$T_1 \times T_2$	9	0	1	−1	−1	$= A_2 + E + T_1 + T_2$

The direct product is now no longer an irreducible representation but a reducible one, which can be reduced in the usual way using the reduction formula (*Chapter 3*). In general:

- if the direct product involves a degenerate system the product itself will be degenerate.

SAQ 11.5 : What is $E_1 \times E_2$ under C_{6v} symmetry ?

Fortunately, a number of trends exist in the nature of these direct products, which allow them to be determined in reduced form without recourse to the reduction formula (Table 11.1).

Ternary direct products can be evaluated as sequential binary direct products. For example, under C_{4v} and using the rules given in Table 11.1 (*overleaf*):

$$B_1 \times B_2 \times E = (B_1 \times B_2) \times E = A_2 \times E = E$$

To confirm this:

C_{4v}	E	$2C_4$	C_2	$2\sigma_v$	$2\sigma_d$	
B_1	1	−1	1	1	−1	
B_2	1	−1	1	−1	1	
$B_1 \times B_2$	1	1	1	−1	−1	$= A_2$
E	2	0	−2	0	0	
$A_2 \times E$	2	0	−2	0	0	$= E$

SAQ 11.6 : Using the rules given in Table 11.1, determine the following direct products:

$C_{2h}: A_g \times A_u \times B_u$

$D_{3h}: A_1'' \times E'' \times A_2'$

$T_d\ : E \times T_1 \times T_2$

$O_h: E_g \times A_{2g} \times T_{1u}$

Table 11.1 Rules for determining direct products.

General rules

$A \times A = A$	$B \times B = A$	$A \times B = B$
$A \times E = E$	$B \times E = E$	$A \times T = T$
$B \times T = T$	$A \times E_1 = E_1$	$A \times E_2 = E_2$
$B \times E_1 = E_2$	$B \times E_2 = E_1$	
Subscripts		
$1 \times 1 = 1$	$1 \times 2 = 2$	$2 \times 2 = 1$
$g \times g = g$	$u \times u = g$	$u \times g = u$
Superscripts		
$' \times ' = '$	$'' \times '' = '$	$' \times '' = ''$

Doubly-degenerate representations

C_3, C_{3h}, C_{3v}, D_3, D_{3h}, D_{3d}, C_6, C_{6h}, C_{6v}, D_6, D_{6h}, S_6, O_h, T_d

$E_1 \times E_1 = E_2 \times E_2 = A_1 + A_2 + E_2$

$E_1 \times E_2 = B_1 + B_2 + E_1$

For C_4, C_{4v}, C_{4h}, D_{2d}, D_4, D_{4h}, S_4

$E \times E = A_1 + A_2 + B_1 + B_2$

(If no subscripts on A, B, or E, read as $A_1 = A_2 = A$ etc.)

Triply-degenerate representations

T_d, O_h

$E \times T_1 = E \times T_2 = T_1 + T_2$

$T_1 \times T_1 = T_2 \times T_2 = A_1 + E + T_1 + T_2$

$T_1 \times T_2 = A_2 + E + T_1 + T_2$

Note that one of the outcomes of these rules is:

- the direct product of a degenerate irreducible representation and itself is a reducible representation which includes the totally symmetric representation

For example, under O_h symmetry $E_g \times E_g = A_{1g} + A_{2g} + E_g$. This is an important point with regard to using selection rules to establish whether or not a transition is symmetry allowed.

We are now in a position to consider electronic transitions which involve degenerate energy levels, as the direct products required to determine the symmetries of the electronic configurations can be made using Table 11.1. As an example, can we predict the electronic spectrum of benzene, at least for the low energy transitions? The MO diagram for C_6H_6 (D_{6h}) was the focus of problem 6 at the end of Chapter 8 (*see Appendix 4 for answer*) and is given in Fig. 11.3 in simplified form.

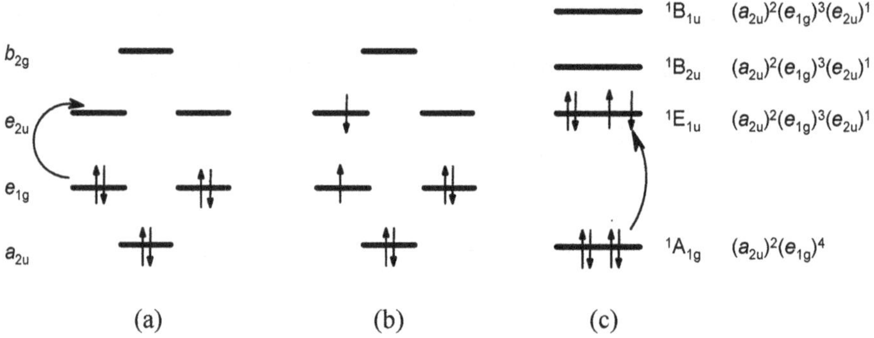

Fig. 11.3 (a) Ground and (b) excited state for C_6H_6; (b) shows only the spin-allowed transition. 11.3(c) shows the only symmetry-allowed transition, from $^1A_{1g} \rightarrow {}^1E_{1u}$ states; the relative energies of the excited states are not significant.

The ground state has electronic configuration $(a_{2u})^2(e_{1g})^4$ which gives rise to a configuration of $^1A_{1g}$ symmetry, while the first excited state has configuration $(a_{2u})^2(e_{1g})^3(e_{2u})^1$. The symmetry label for this latter configuration is given by the direct product of the terms for each orbital. For the components $(a_{2u})^2$ (= A_{1g}, as it is a full orbital) and $(e_{2u})^1$ (= E_{2u}, a singly-occupied orbital has the same symmetry as the orbital itself) the situations are straightforward, but a problem arises for the symmetry of the $(e_{1g})^3$ level as $E_{1g} \times E_{1g} \times E_{1g}$ is an instance where using a direct product method does not give the correct symmetry. At first thought, the presence of more than one electron to distribute among the orbitals might be expected to give rise to more permutations, and this is what taking the triple direct product would deliver. However, adding more electrons starts to fill orbitals, which, by the Pauli principle, can't take more than two electrons, so the number of permutations of electrons in orbitals actually starts to fall when the available orbitals become more than half filled. In these cases [as in the current case for $(e_{1g})^3$] the easiest way to determine the symmetry of the configuration is by a **hole formalism**. By this we mean that if m electrons occupy n available equivalent locations, then electronic configurations $(m)e$ and $(n\text{-}m)e$ are equivalent e.g. if a doubly-degenerate orbital holds only three (not four) electrons, then (4 - 3)e ≡ 1e. In this case, instead of considering the orbitals to hold three electrons we can think of there being one hole present, and one hole ≡ one electron in terms of symmetry. On this basis, the $(e_{1g})^3$ configuration is the same as $(e_{1g})^1$ which has E_{1g} symmetry.

The symmetry of the $(a_{2u})^2(e_{1g})^3(e_{2u})^1$ configuration is, therefore, given by:

$$A_{1g} \times E_{1g} \times E_{2u} = E_{1g} \times E_{2u} = B_{1u} + B_{2u} + E_{1u}$$

This is reasonable, as the four $(e_{1g})^3(e_{2u})^1$ electrons could be arranged four ways (a hole in either of the degenerate e_{1g} pair and an electron in either of the two e_{2u} orbitals), and we have symmetry labels which correspond to this number of configurations.

To decide which, if any, of these excited states are of the correct symmetry to make the transition symmetry allowed, we need to find out if any of the following direct products contain the totally symmetric A_{1g} label:

$$A_{1g} \times \begin{pmatrix} \mu_x \\ \mu_y \\ \mu_z \end{pmatrix} \times \begin{pmatrix} B_{1u} \\ B_{2u} \\ E_{1u} \end{pmatrix} = A_{1g} \times \begin{pmatrix} A_{2u} \\ E_{1u} \end{pmatrix} \times \begin{pmatrix} B_{1u} \\ B_{2u} \\ E_{1u} \end{pmatrix}$$

$$\begin{aligned} A_{1g} \times A_{2u} \times B_{1u} &= B_{2g} \\ A_{1g} \times A_{2u} \times B_{2u} &= B_{1g} \\ A_{1g} \times A_{2u} \times E_{1u} &= E_{2g} \\ A_{1g} \times E_{1u} \times B_{1u} &= E_{1g} \\ A_{1g} \times E_{1u} \times B_{2u} &= E_{2g} \\ A_{1g} \times E_{1u} \times E_{1u} &= A_{1g} + A_{2g} + E_{2g} \end{aligned}$$

The only allowed transition is that from the ground state $(^1A_{1g})$ to the excited state arrangement which has $^1E_{1u}$ symmetry, as it is the only one which contains the label A_{1g} in its transition integral (Fig. 11.3c).

Note that we could reduce the amount of work involved in this process by applying the shortcut mentioned at the end of Section 11.2, by re-ordering the ternary direct product $\psi_i \times \mu \times \psi_f$ as $\psi_i \times \psi_f \times \mu$. For example, in the following two cases from the example of benzene, above, we can focus primarily on the direct products of terms in bold $(\psi_i \times \psi_f)$:

$$\mathbf{A_{1g}} \times E_{1u} \times \mathbf{E_{1u}}$$

Here $\mathbf{A_{1g}} \times \mathbf{E_{1u}} = E_{1u}$ and since $E_{1u} \times E_{1u}$ is the product of two identical representations the product must contain A_{1g} and the transition is symmetry-allowed. Conversely:

$$\mathbf{A_{1g}} \times A_{2u} \times \mathbf{E_{1u}}$$

Now, $\mathbf{A_{1g}} \times \mathbf{E_{1u}} = E_{1u}$ and since $A_{2u} \times E_{1u}$ is not the product of identical representations the outcome can't contain A_{1g} so the transition is symmetry-forbidden.

In other words, the simple strategy for deciding if any transition is symmetry-allowed or not is :

- take the binary direct product $\psi_i \times \psi_f$. If it has the same symmetry as any component of μ then the transition is allowed, if not it is forbidden.

SAQ 11.7 : Which of the following transitions are allowed under O_h symmetry ?
$A_{1u} \rightarrow T_{2g}$
$E_u \rightarrow T_{1g}$

No mention has been made so far in this analysis of degenerate systems of the spin of the electrons. As the electrons in the e_{1g} and e_{2u} levels can have spin of either ½ or -½, then both singlet and triplet variations on the configurations B_{1u} + B_{2u} + E_{1u} are possible. Since the ground state is a spin singlet, only transitions to singlet excited states are allowed, as shown in Fig. 11. 2(b). The three transitions which go from the singlet $^1A_{1g}$ ground state to any of the triplet excited states e.g. $^1A_{1g} \rightarrow {}^3E_{1u}$, that is in which the electron flips its spin on promotion, are spin forbidden.

11.5 EPILOGUE – SELECTION RULES FOR VIBRATIONAL SPECTROSCOPY

In Chapter 4 we stated the selection rule for an infrared-active mode simply as:

- a vibrational mode is infrared active if it has the same symmetry as one of the translational vectors ($\mathbf{T_x}$, $\mathbf{T_y}$ or $\mathbf{T_z}$), read from the character table.

Some justification for this statement is now in order.

The absorption of infrared radiation leading to the excitation of vibrational modes within a molecule is also mediated by the electric dipole within the molecule, just as in electronic spectroscopy. As such, equation 11.1 is equally applicable to infrared spectroscopy, and the same selection rule – that for an allowed transition the integral must be non-zero i.e. the direct product must contain the totally symmetric irreducible representation for the point group – also applies. The ground vibrational state (ψ_i) is always totally symmetric (a proof is beyond the scope of this text), so for the direct product to be, at least in part, totally symmetric then one of the following direct products must be totally symmetric:

$$\psi_i \times \begin{pmatrix} \mathbf{T_x} \\ \mathbf{T_y} \\ \mathbf{T_z} \end{pmatrix} \times \psi_f \quad = \quad \text{"A"} \times \begin{pmatrix} \mathbf{T_x} \\ \mathbf{T_y} \\ \mathbf{T_z} \end{pmatrix} \times \psi_f$$

where "A" corresponds to the totally symmetric representation appropriate to the point group. Since the direct product of any function with "A" leaves it unchanged (Table 11.1), the triple direct products shown above will contain "A" only if the one of the following generates "A":

$$\mathbf{T_x} \times \psi_f$$
$$\mathbf{T_y} \times \psi_f$$
$$\mathbf{T_z} \times \psi_f$$

This will only happen if the symmetry of ψ_f includes at least one of $\mathbf{T_x}$, $\mathbf{T_y}$ or $\mathbf{T_z}$ symmetry labels, as mentioned previously in Section 11.2.

The situation with the Raman selection rule is analogous, if a little more complex:

- a vibrational mode is Raman active if it has the same symmetry as one of the binary combinations (xy, xz, x^2-y^2 etc), read from the character table.

Excitation of a vibrational mode in a Raman experiment requires a change in polarisability within the molecule, and the following integral needs to be non-zero:

$$\int \psi_i \alpha \psi_f \ d\tau \qquad \text{(eqn 11.2)}$$

α is the molecule's polarisability and is a tensor, a 3 x 3 matrix of elements α_{jk} i.e. α_{x^2}, α_{xy}, α_{xz} etc, where x, y, z are the rows and columns of the matrix. In short, the symmetry of these tensor components is that given by the binary combinations listed on the right of the character table (e.g. α_{xy} has the same symmetry as xy), so provided ψ_f has the same symmetry as one of these (x^2, y^2, z^2, xy, xz, yz or their combinations) then the direct product $\alpha_{jk} \times \psi_f$ will be totally symmetric and one of the ternary direct products "A" $\times$ $\alpha_{jk} \times \psi_f$ will also be non-zero.

11.6 SUMMARY

- a singly-occupied orbital has the same symmetry as the orbital itself.
- any filled orbital (or set of degenerate orbitals) is totally symmetric (top row of the character table).
- the direct product of a non-degenerate irreducible representation and itself is always the totally symmetric representation.
- the direct product of a degenerate irreducible representation and itself is a reducible representation which contains the totally symmetric representation.
- the direct product of two different irreducible representations is never the totally symmetric representation.
- if the binary direct product $\psi_i \times \psi_f$ has the same symmetry as μ the transition is allowed, if not it is forbidden.
- for a spin-allowed transition $\Delta S = 0$ i.e. there is no change in electron spin.

PROBLEMS

Answers to all problems marked with * are given in Appendix 4.

1*. Based on the MO diagram for NH_3 (Fig. 8.5), predict which of the transitions of an electron in the HOMO to any of the unoccupied MOs is symmetry-allowed. Include spin multiplicities in all symmetry labels.

2. Repeat question 1 to predict the allowed low-energy electronic transitions for BH_3 (Fig. 8.4).

3*. *Trans*-butadiene (C_{2h}) has the following MO diagram for the π-bond framework:

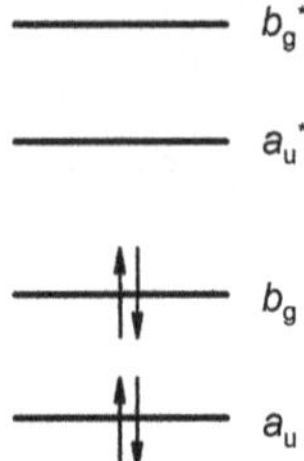

Are excitations from the HOMO to either of the unoccupied π* MOs symmetry allowed ?

4*. The cyclobutadiene dianion $[C_4H_4]^{2-}$ (D_{4h}) has a π-bond framework of a_{2u}, e_g, and b_{2u}* symmetries, in order of increasing MO energy; the six π-electrons occupy the a_{2u} and e_g MOs.
Are either of the transitions from the filled a_{2u}, e_g MOs to the LUMO symmetry allowed ?

12

TERMS AND CONFIGURATIONS

The chemistry of the transition metals is dominated by the presence of partly filled *d*-orbitals. Not only does this give rise to variable oxidation states, but key properties such as magnetism and colour are also directly related to the distribution of electrons among the available orbitals. In this and the following chapter we will explore the role of symmetry in understanding the electronic spectra of transition metal compounds. This is a complex area and the aim of these chapters is to give insights into how symmetry can be used to explain the origins of the observed spectra. It will not attempt to rationalise these spectra completely, as issues beyond symmetry become important and these are outside the scope of this text. Group theory can only provide a qualitative – but not quantitative – understanding of the background to electronic transitions.

For simplicity, only perfectly octahedral and tetrahedral complexes will be considered, restricting discussion to the O_h and T_d point groups. The orbitals involved are the metal-centred t_{2g} and e_g* (O_h) or t_2* and e (T_d), as described earlier (Fig. 9.7, *SAQ 9.3*), though for clarity of presentation in this chapter we will simplify these to either t_{2g} / e_g or t_2 / e, as in simple Crystal Field Theory (CFT). The approach will be to explain how symmetry labels can be used to describe the arrangement of d^n electrons across these available orbitals in both ground and excited states, in the same way that symmetry labels for the electron distributions across MOs in benzene were derived in Chapter 11. These d^n arrangements are known as **configurations** and can be broken down into a series of **microstates** in which the n electrons are distributed across different *d*-orbitals. Those microstates for a d^n configuration with the similar energy[†] can be grouped together and described collectively by a **term symbol**.

In our analysis, we will need to:

[†] Microstates described by the same term symbol all have the same orbital and spin angular momenta but become different in energy when these are coupled together (*see* Fig. 12.1).

- determine the symmetry labels for all arrangements of electrons for a given d^n configuration, which will cover ground and excited states for this configuration.
- identify the spin multiplicities of these states, since transitions between states of the same multiplicity are responsible for the most intense bands in the spectra.
- chose the group of states (terms) of the same multiplicity and which include the lowest energy (ground state) term, as these will form the basis of the spin-allowed electronic transitions.

The focus of the discussion will be octahedral complexes, and only when the relevant methodology has been covered and simplifications identified will tetrahedral complexes be considered. Before this, however, it is important to establish some fundamental descriptions of electronic states and how these behave in a ligand environment.

12.1 TERM SYMBOLS

Consider any d^n configuration. If the there are no ligands surrounding the metal (**a free ion**), all five *d*-orbitals are equal in energy and the *n* electrons can be accommodated in any of them, with spin of either ½ or -½. Thus there are several arrangements, or **microstates**, for these *n* electrons, some of which have the same energy as each other, others of which are different due to less favourable *e-e* interactions. The number of such possible arrangements is given by the formula:

$$D_t = N! / (N_e)!(N_h)! \qquad \text{(eqn 12.1)}$$

D_t = total number of microstates (total degeneracy of the configuration)
N = number of spin orbitals (i.e. 2 × no. orbitals)
N_e = number of electrons
N_h = number of holes
N! means "N factorial" and is 1 × 2 × 3…× N

For a d^1 configuration, N = 2 × 5, N_e = 1, N_h = 9 so :

$$D_t = (2 \times 5)! / 1! \times 9! = 10! / 1! \times 9! = 10$$

That is, there are 10 ways in which one electron can be placed into any one of five equal-energy orbitals with either of two possible spins.

SAQ 12.1 : How many microstates are there for a d^2 configuration? .
Answers to all SAQs are given in Appendix 3.

The number of microstates increases up to d^5, then decreases as the permutations are limited by the fact that each of the orbitals is already partly filled. Table 12.1 shows this relationship, and also indicates that d^n and $d^{(10-n)}$ configurations are related, a fact we shall make use of as the topic unfolds.

Table 12.1 Microstates for various d^n configurations.

d^n	d^1	d^2	d^3	d^4	d^5	d^6	d^7	d^8	d^9	d^{10}
D_t	10	45	120	210	252	210	120	45	10	1

Are all the microstates of a d^n configuration equal in energy? The answer is no, because, among several factors, two electrons in the same orbital repel each other more than two electrons in separate orbitals. The microstates divide into several groups, in which each group differs in energy from the others, but within each group all the microstates have the same orbital and spin angular momenta. How these microstates are determined and placed into groups is exemplified briefly in Appendix 2, and is dealt with in depth in many undergraduate level texts on inorganic chemistry.

As an example, for d^2 the 45 microstates are divided as follows, with the number of microstates in each group given in parentheses :

$$d^2 = {}^1S\,(1),\ {}^1D\,(5),\ {}^1G\,(9),\ {}^3P\,(9),\ {}^3F\,(21)$$

Each group of microstates is labelled by a **term symbol**; these symbols describe multi-electron arrangements associated with a set of quantum numbers which exactly parallel the nomenclature for single-electron cases i.e. atomic orbitals. Each term (S, P, D, F, G) corresponds to a maximum orbital quantum number (L, the sum of the m_l quantum numbers for individual electrons) in which the term symbols map onto L in the same way that the labels for atomic orbitals correspond to l values ($l = 0$ is s, $l = 1$ is p etc). Each L has $2L + 1$ associated M_L values, just has l has $2l + 1$ m_l values e.g. $l = 1$ has $m_l = 1, 0, -1$ which relate to p_x, p_y, p_z. The spin multiplicity (singlet, triplet etc), the left hand superscript in the term symbol, has been described earlier in Section 11.3.

Table 12.2 Parallels in nomenclature between single-electron wavefunctions (atomic orbitals) and multi-electron terms.

orbital	l	m_l	**term**	L	M_L	degeneracy
s	0	0	S	0	0	1
p	1	1, 0, -1	P	1	1, 0, -1	3
d	2	2, 1, 0, -1, -2	D	2	2, 1, 0, -1, -2	5
f	3	3, 2, 1, 0, -1, -2, -3	F	3	3, 2, 1, 0, -1, -2, -3	7
g	4	4, 3,..... 0,....-3, -4	G	4	4, 3,.....0,....-3, -4	9

The total degeneracy of each term (i.e. the number of microstates it represents) is given by the product of the spin degeneracy ($2S + 1$) and the orbital degeneracy ($2L + 1$), and these are given in parentheses (above) for the microstates associated with d^2. Note the total number of microstates for d^2 sums to 45, as required.

What are the relative energies of these terms? Here we make use of Hund's rules, which states that the lowest energy requires electrons to be as far apart as possible within degenerate orbitals to minimise $e - e$ repulsions (a **coulombic energy** term), and keep spins parallel for as long as possible (this maximises what is known as the

exchange energy). These two effects are commonly grouped together and termed a **pairing energy**. In quantum mechanical terms, these ideas can be re-stated as:

- the lowest energy (ground state) has maximum S (most important).
- for terms of the same S, the lowest energy has maximum L.

In the case of the d^2configuration, the lowest energy set of microstates has term symbol ^3F, triplets being lower in energy than singlets (maximum S), and F being lower in energy than P (maximum L). Since the spin selection rule only allows transitions between terms of the same spin multiplicity, our prime concern is with terms of the same total spin as the ground term; for d^2 these are ^3F (ground) and ^3P (excited).

For completeness, the terms associated with the free ions for all the d^n configurations are listed in Table 12.3, with the ground state term highlighted in bold. Note the relationship between d^n and $d^{(10\text{-}n)}$ configurations, and that only $d^2(d^8)$ and $d^3(d^7)$ have excited states of the same multiplicity as the free ion.

Table 12.3 Free-ion terms for d^n configurations

d^n	Free-ion terms[a]
d^0, d^{10}	$^1\mathbf{S}$
d^1, d^9	$^2\mathbf{D}$
d^2, d^8	$^1S, {}^1D, {}^1G, {}^3P, {}^3\mathbf{F}$
d^3, d^7	$^2P, {}^2D(2), {}^2F, {}^2G, {}^2H, {}^4P, {}^4\mathbf{F}$
d^4, d^6	$^1S(2), {}^1D(2), {}^1F, {}^1G(2), {}^1I, {}^3P(2), {}^3D, {}^3F(2), {}^3G, {}^3H, {}^5\mathbf{D}$
d^5	$^2S, {}^2P, {}^2D(3), {}^2F(2), {}^2G(2), {}^2H, {}^2I, {}^4P, {}^4D, {}^4F, {}^4G, {}^6\mathbf{S}$

[a] The number of times a term arises is given in parentheses

To summarise :

- a given electronic configuration is split into terms of different multiplicity based on whether spins align or oppose. This is called coupling of spin angular momenta. The lowest energy equates to the maximum number of aligned spins i.e. maximum S.
- the configuration is further split depending on which orbitals are occupied and the implications of this for e - e repulsions. This is called coupling of orbital angular momenta. Energy is lowest for maximum L.
- finally, the effects of spin and orbital motion are not indepenedent but couple together (spin-orbit coupling); this separates the energies of microstates within a term.

Figure 12.1 (*overleaf*) puts these effects together in order of relative importance in a schematic form for the d^2 configuration. The right subscript in the term symbol is the J value arising from spin-orbit coupling and takes values from $L+S$ to $L-S$ under the **Russell-Saunders** scheme. The magnitude of spin-orbit coupling is small compared to that of the e - e interactions and will be largely ignored in our analysis of *d-d* spectra, though it is of relevance when discussing violation of the spin selection rule (*Section 13.3*). † The two types of *e-e* interactions (coupling of spin, coupling of orbital angular momenta) are of similar magnitude, so while Hund's rules identify the term of lowest energy, the relative energies of other terms require detailed calculations; for d^2 (Fig. 12.1), the ^3P term turns out to be higher in energy than ^1G despite its higher multiplicity.

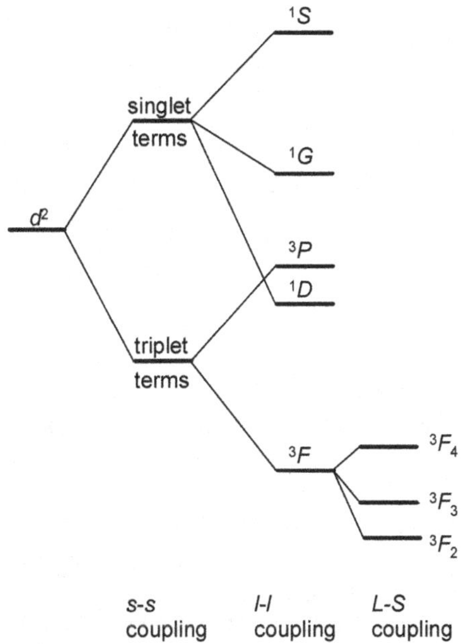

Fig. 12.1 Splitting of terms for multi-electron systems, exemplified by d^2 (*L-S* coupling is shown only for the ground term).

12.2 THE EFFECT OF A LIGAND FIELD – ORBITALS

The parallel between terms and atomic orbitals shown in Table 12.2 suggests it is worthwhile considering how various orbitals behave under different symmetries, as a guide to how related terms of a free ion might be influenced under the same "real world" conditions of a ligand environment. We have previously stated, without qualification (*Sections 7.2 and 9.2*), that:

- *s*-orbitals are totally symmetric and have a symmetry label corresponding to the top row of the relevant character table.
- *p*-orbitals have the same symmetry as $\mathbf{T_x}$, $\mathbf{T_y}$ and $\mathbf{T_z}$.

† This only applies to first row transition elements. Heavier *d*-block metals have more significant spin-orbit coupling and its effect on electronic spectra is more pronounced. Furthermore, an alternative coupling mechanism (known as *j-j* coupling) is more appropriate than the Russell-Saunders scheme.

- *d*-orbitals have the same symmetry as the corresponding binary function, as read from the character table.

Some detail is now required to justify these statements. Just as we were able to write expressions for $\chi_{u.a.}$ to help determine the reducible representation for a collection of atom displacements (*Section 3.4*), similar, though more complex, expressions are available for orbitals on unshifted atoms. These are not derived but are simply stated:

$$\chi[E] = 2j + 1 \qquad \text{(eqn. 12. 2)}$$

$$\chi[C_n] = \sin[(j + ½)\theta] / \sin(\theta/2) \qquad \text{(eqn. 12. 3)}$$

$$\chi[i] = \pm(2j + 1) \qquad \text{(eqn. 12. 4)}$$

$$\chi[S_n] = \pm \sin[(j + ½)(180 + \theta)] / \sin[(180 + \theta)/2] \qquad \text{(eqn. 12. 5)}$$

$$\chi[\sigma] = \pm \sin[(j + ½)180] \qquad \text{(eqn. 12. 6)}$$

where $\theta = (360/n)$. Two notes of qualification go with these equations. Firstly, j here is a general quantum number that can be varied, and should not be confused with the J quantum number which forms part of the spin-orbit coupling scheme. The equations can be applied to the determination of the reducible representations for orbitals by using (for j) l (the angular momentum quantum number), for terms (using L) or spin states (using S). Secondly, equations 12.4 – 12.6 include a "±", which varies according to the function under consideration. If the function is g, such as any of the *d*-orbitals, then + is used; if it is u e.g. *p*- or *f*-orbitals, then use −. Terms are more difficult to generalize about as it depends on their origin. For example, a single electron in an *f*-orbital gives rise to a ^2F term, and so both the *f*-orbital and the associated term are u. Conversely, the ^3F term which arises from a d^2 configuration (Table 12.3) is g because the two *d*-electrons which generate this term have g symmetry.

Table 12.4 χ per group of orbitals on unshifted atoms and related terms arising from a d^n configuration. [a]

	p-orbital	*P* term	*d*-orbital	*D* term	*f*-orbital	*F* term
E	3	3	5	5	7	7
C_2	−1	−1	1	1	−1	−1
C_3	0	0	−1	−1	1	1
C_4	1	1	−1	−1	−1	−1
i	−3	3	5	5	−7	7
S_4	−1	1	−1	−1	1	−1
S_6	0	0	−1	−1	−1	1
σ	1	−1	1	1	1	−1

[a] For example, the three *p*-orbitals together contribute a character -1 for a C_2 rotation for any given point group.

As we are only interested in transition metal complexes, all terms arising from a d^n configuration will have g symmetry. For convenience, χ for key symmetry operations applied to *p*-, *d*-, *f*-orbitals along with terms arising from a d^n configuration (*P*, *D*, *F*) have been evaluated and are listed in Table 12.4 (*above*).

Using the data in Table 12.4 we can demonstrate easily how the five *d*-orbitals split under O_h symmetry:

O_h	E	$8C_3$	$6C_2$	$6C_4$	$3C_2{}^a$	i	$6S_4$	$8S_6$	$3\sigma_h$	$6\sigma_d$	
$\Gamma_{d\text{-orbitals}}$	5	−1	1	−1	1	5	−1	−1	1	1	$= t_{2g} + e_g$

> *SAQ 12.2 : Use Table 12.4 to determine the symmetries of the three p-orbitals under C_{2v} symmetry.*

12.3 SYMMETRY LABELS FOR d^n CONFIGURATIONS – AN OPENING

In an octahedral complex the energies of the five *d*-orbitals are not equivalent, but split into two groups, a lower energy set of three orbitals of t_{2g} symmetry and a higher energy e_g pair. For any d^n configuration a number of arrangements of electrons between these orbitals are possible each of which will have a different energy. Furthermore, for octahedral complexes with $d^4 - d^7$ configurations there are two possible lowest energy situations and which is adopted as the ground state depends crucially on the relative magnitudes of Δ_o and the **pairing energy**. Both **low-spin** and **high-spin** arrangements are possible, as exemplified by d^5:

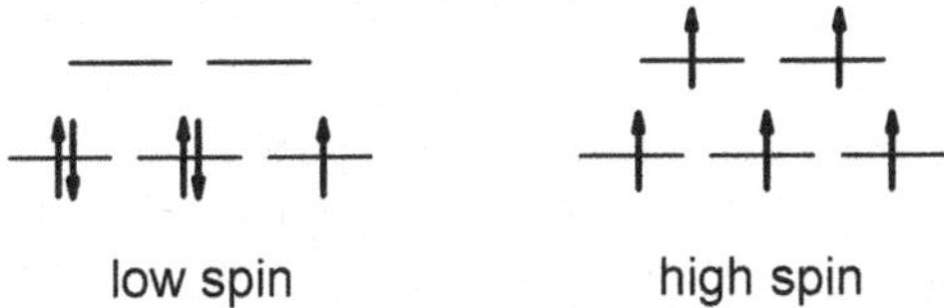

Low-spin complexes occur in **strong ligand fields** (large Δ_o) as the energy required to overcome $e - e$ repulsions when electrons are paired in the same orbital is less than Δ_o. Conversely, when Δ_o is small, a **weak ligand field**, placing electrons in an orbital of only marginally higher energy becomes more favourable and high-spin complexes arise. For tetrahedral complexes the situation is simplified by the fact that Δ_t is small and always equates to a weak field; tetrahedral complexes are always high-spin.

Bearing this in mind we now need to determine the symmetry labels for all the possible configurations for a set of *n* *d*-electrons. This will be a two stage process. First, we need the symmetry labels themselves, then we will need to establish the spin multiplicity associated with each label.

Based on the technique of direct products outlined in Chapter 11, we can evaluate the symmetry label for any d^n configuration for a complex of O_h symmetry. Points to note by way of revision are:

- a singly-occupied orbital has the same symmetry as the orbital itself e.g. $(t_{2g})^1 = T_{2g}$, $(e_g)^1 = E_g$.

- any filled orbital or set of degenerate orbitals is totally symmetric e.g. $(t_{2g})^6 = A_{1g}$, $(e_g)^4 = A_{1g}$.
- for degenerate orbitals of total capacity *n* electrons but only occupied by *m* electrons, (*m*)*e* and (*n-m*)*e* configurations are equivalent by the hole formalism, e.g. $(t_{2g})^1 \equiv (t_{2g})^5 = T_{2g}$; $(e_g)^1 \equiv (e_g)^3 = E_g$. This is of importance for orbitals which are more than half-filled.

These points cover most of the d^n configurations, but the cases of $(t_{2g})^2$, $(t_{2g})^3$ and $(e_g)^2$ need further explanation: The symmetry labels for the $(t_{2g})^2$ and $(e_g)^2$ configurations are obtained by taking direct products and using the rules given in Table 11.1. Since we will be using these rules frequently in this chapter, the key products relating to the O_h point group are summarised in the following table; subscripts *g*, *u* need to be added as appropriate using the rules of Table 11.1

Table 12.5 Direct product table for octahedral symmetry

	A_1	A_2	E	T_1	T_2
A_1	A_1	A_2	E	T_1	T_2
A_2	A_2	A_1	E	T_2	T_1
E	E	E	$A_1 + A_2 + E$	$T_1 + T_2$	$T_1 + T_2$
T_1	T_1	T_2	$T_1 + T_2$	$A_1 + E + T_1 + T_2$	$A_2 + E + T_1 + T_2$
T_2	T_2	T_1	$T_1 + T_2$	$A_2 + E + T_1 + T_2$	$A_1 + E + T_1 + T_2$

The direct product for $(t_{2g})^2$, and thus also $(t_{2g})^4$ by the hole formalism, is:

$$T_{2g} \times T_{2g} = A_{1g} + E_g + T_{1g} + T_{2g}$$

Note that at this stage we are making no assignments for the multiplicities of these **ligand-field terms** (terms arising from a d^n configuration in an environment of ligands). All we know at this stage is that, using eqn 12.1, there are 15 microstates associated with a $(t_{2g})^2$ configuration ($N = 6$, $N_e = 2$, $N_h = 4$), some will be triplets (spins parallel) and some singlets (spins paired). Note also that the number of microstates here (15) differ from the total microstates associated with d^2 (45) since we are only dealing with electrons in the three t_{2g}, not all five, *d*-orbitals.

SAQ 12.3 : What are the ligand field terms for the $(e_g)^2$ configuration, ignoring spin multiplicities? What is the degeneracy of this configuration?

The most problematic group of ligand-field terms to evaluate is for $(t_{2g})^3$, as this is a case when taking the direct products [$T_{2g} \times T_{2g} \times T_{2g} = (A_{1g} + E_g + T_{1g} + T_{2g}) \times T_{2g}$ = etc] leads to the wrong answer, for the same reason as decribed for $(e_{1g})^3$ in benzene in Section 11.4. Unfortunately, unlike this latter case where we could invoke the hole formalism to help, this is not possible for $(t_{2g})^3$ [(*n-m*)*e* and (*m*)*e* arrangements are the same for $n = 6$, $m = 3$] so another approach is needed. We must avoid putting all three electrons into one orbital (Pauli principle), and one way to achieve this to take an arrangement for $(t_{2g})^2$ which keeps both electrons in separate orbitals, so that when a third electron is added it cannot place all three in the same location. If we ignore spin, there are three ways of putting the two electrons (or one hole!) of a $(t_{2g})^2$ configuration in three orbitals, so these must be described by either the T_{1g} or T_{2g} labels and not A_{1g} or E_g. We can distinguish which of T_{1g} or T_{2g} is correct by a consideration of spin multiplicity, but we are not yet at the stage where this distinction can be made. It turns out to be the T_{1g} term (a fact we will justify in Section 12.4), so the relevant direct product for $(t_{2g})^3$ is, bearing in mind that an electron in a singly-occupied orbital has the same symmetry as that orbital:

$$(t_{2g})^2 \times (t_{2g})^1 = T_{1g} \times T_{2g} = A_{2g} + E_g + T_{1g} + T_{2g}$$

The results of all these analyses, still lacking any comment on spin multiplicities, are summarised in Table 12.6.

Table 12.6 Configurations and associated term symbol symmetry labels

config^n	micro-states	Term symbols	config^n	micro-states	Term symbols
$(t_{2g})^1$	6	T_{2g}	$(e_g)^1$	4	E_g
$(t_{2g})^2$	15	$A_{1g} + E_g + T_{1g} + T_{2g}$	$(e_g)^2$	6	$A_{1g} + A_{2g} + E_g$
$(t_{2g})^3$	20	$A_{2g} + E_g + T_{1g} + T_{2g}$	$(e_g)^3$	4	E_g
$(t_{2g})^4$	15	$A_{1g} + E_g + T_{1g} + T_{2g}$	$(e_g)^4$	1	A_{1g}
$(t_{2g})^5$	6	T_{2g}			
$(t_{2g})^6$	1	A_{1g}			

The next task is to consider the spin multiplicities of the terms in this table, some of which are straightforward, other less so. The $(t_{2g})^1$ and $(e_g)^1$ configurations can only be spin doublets $(2S + 1)$, so the term symbols are $^2T_{2g}$ [as is $(t_{2g})^5$] and ^2E_g [as is $(e_g)^3$], respectively. $(t_{2g})^6$ and $(e_g)^4$ configurations both describe filled orbitals with all spins paired so must be singlet terms i.e. both are $^1A_{1g}$.

$(t_{2g})^3$ is also straightforward, though a little more thought is needed here. We know that the maximum multiplicity for 3 electrons is a quartet ($S = {}^3/_2$) and that the $A_{2g} + E_g + T_{1g} + T_{2g}$ terms must generate 20 microstates. The only way three electrons can have the parallel spins that generate the quartet multiplicity is with one in each of the three t_{2g} orbitals, and as this can only be done one way it must equate to an A term i.e. $^4A_{2g}$. If there is any pairing among the three electrons we are left with a doublet state, as the spins paired have $S = 0$ so the odd electron gives $S = {}^1/_2$. There is only one combination that generates the required 20 microstates and, bearing in mind that the number of microstates for a given term is given by the product of its spin and orbital multiplicities, that is:

$$(t_{2g})^3 = {}^4A_{2g} + {}^2E_g + {}^2T_{1g} + {}^2T_{2g} \qquad (= 4 + 4 + 6 + 6 = 20 \text{ microstates})$$

More problematic to deal with are the multiplicities of the terms associated with $(t_{2g})^2$ [and thus also $(t_{2g})^4$] and $(e_g)^2$ arrangements. We can get some guidance from the number of microstates associated with each configuration, as we did in the case of $(t_{2g})^3$. For example, for $(e_g)^2$ there are 6 associated microstates, and the two electrons could be parallel (triplet multiplicity) or paired (singlet). So:

$$(e_g)^2 = {}^{1,3}A_{1g} + {}^{1,3}A_{2g} + {}^{1,3}E_g$$

We can rule out a ^3E_g term, as this alone would account for the 6 microstates, so we must have:

$$(e_g)^2 = {}^1A_{1g} + {}^3A_{2g} + {}^1E_g \quad \text{or} \quad {}^3A_{1g} + {}^1A_{2g} + {}^1E_g$$

That is, one A term is a singlet and the other a triplet (to give 6 microstates in total), but we don't know which way around this is.

For $(t_{2g})^2$, where both triplet and singlet possibilities can exist among the 15 microstates i.e the two electrons could be parallel or paired, we have:

$$(t_{2g})^2 = {}^{1,3}A_{1g} + {}^{1,3}E_g + {}^{1,3}T_{1g} + {}^{1,3}T_{2g}$$

Three possibilities for the 15 microstates exist:

$${}^1A_{1g} + {}^1E_g + {}^1T_{1g} + {}^3T_{2g} \ : \ (1+2+3+9 \text{ microstates})$$

$${}^1A_{1g} + {}^1E_g + {}^3T_{1g} + {}^1T_{2g} \ : \ (1+2+9+3 \text{ microstates})$$

$${}^3A_{1g} + {}^3E_g + {}^1T_{1g} + {}^1T_{2g} \ : \ (3+6+3+3 \text{ microstates})$$

The combination which includes the $^3A_{1g}$ term can be eliminated, as the arrangement of two electrons with parallel spins across three *d*-orbitals must occur in three ways and thus be a T term:

In this case, then, we are left with deciding which of the terms T_{1g} or T_{2g} is the spin triplet.

To resolve these remaining ambiguities for $(t_{2g})^2$ and $(e_g)^2$ we need to take a step back to the free ion case for guidance.

12.4 WEAK LIGAND FIELDS, TERMS AND CORRELATION DIAGRAMS

In the absence of any ligands, all five *d*-orbitals are degenerate and we have established the term symbols which are associated with groups of microstates which arise from the different arrangements of electrons within these orbitals; based on maximum spin multiplicity, the lowest energy group(s) of microstates (terms) have been established (Section 12.1, Table 12.3). When a group of ligands surrounds the central ion the symmetry of the system is lowered and the degeneracy of the orbitals – and hence microstates – is, at least partially, lifted. We are interested to begin with with the effect of a weak ligand field i.e. a very small Δ, as this will lift the orbital degeneracy without changing the criterion that the lowest energy term(s) will equate to maximum multiplicity. In other words, the multiplicity of terms associated with the free ion will be unaffected. Of course, as Δ increases, some of the microstates associated with maximum multiplicity will increase in energy as the energy of the orbitals in which some of them reside increases and so the picture will change. For example, for the d^2 case an electron in each of d_{xy} and d_{z^2} is contained in the set of microstates of maximum multiplicity for the free ion but which become increasingly higher in energy as Δ increases, i.e. for an octahedral complex:

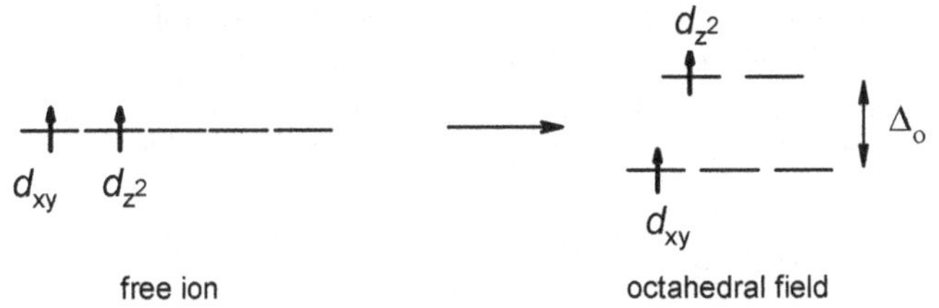

d_{z^2}

d_{xy} d_{z^2} → Δ_o

d_{xy}

free ion octahedral field

The effect of a weak ligand field on the various free-ion terms can be determined by application of equations 12.2 – 12.6.

O_h	E	$8C_3$	$6C_2$	$6C_4$	$3C_2^{a}$	i	$6S_4$	$8S_6$	$3\sigma_h$	$6\sigma_d$	
Γ_S	1	1	1	1	1	1	1	1	1	1	$= A_{1g}$
Γ_P	3	0	−1	1	−1	3	1	0	−1	−1	$= T_{1g}$
Γ_D	5	−1	1	−1	1	5	−1	−1	1	1	$= E_g + T_{2g}$
Γ_F	7	1	−1	−1	−1	7	−1	1	−1	−1	$= A_{2g} + T_{1g} + T_{2g}$

Note that the symmetry labels for all the terms resulting for the effect of the ligand field on the free-ion term are *g*, a feature which will become important when considering selection rules for $d - d$ transitions.

The spin multiplicities have been omitted from the symmetry labels derived above, but are the same as for the free-ion term. We can make the generalisation that:

- the ligand-field terms have the same spin multiplicity as the free-ion term from which they are derived.

The ground-state term arising from a d^2 configuration (^3F; Table 12.3) thus splits into $^3A_{2g}$ + $^3T_{1g}$ + $^3T_{2g}$. Furthermore, as would be expected, these labels cover the same 21 microstates that correspond to the ^3F free-ion term, determined by combining the spin multiplicity with the degeneracy of the symmetry label:

$^3A_{2g}$ = 3 (spin multiplicity) × 1 (A label) = 3
$^3T_{1g}$ = 3 × 3 = 9
$^3T_{2g}$ = 3 × 3 = 9

SAQ 12.4 : How many microstates are associated with the ^4F ground term for a d^3 configuration ? How are these split in an octahedral field ?

Using the same approach we can determine how the free-ion terms split in other weak ligand fields, the most important of which is T_d. This is a point that we will return to in Section 13.4.

SAQ 12.5 : Use equations 12.2 – 12.6 to determine a reducible representation for the ^1G term associated with a d^2 configuration in an octahedral field and then reduce this to identify the terms arising from this splitting.

The problems remaining from Section 12.3 are the multiplicities of terms for the $(t_{2g})^2$ and $(e_g)^2$ configurations, both of which can be linked to a d^2 system, so this now becomes the focus of the remainder of this section.

We have seen above how the terms for the d^2 free ion (Table 12.3) split in a weak octahedral field:

$$^1S = {}^1A_{1g}$$
$$^3P = {}^3T_{1g}$$
$$^1D = {}^1E_g + {}^1T_{2g}$$
$$^3F = {}^3A_{2g} + {}^3T_{1g} + {}^3T_{2g}$$
$$^1G = {}^1A_{1g} + {}^1E_g + {}^1T_{1g} + {}^1T_{2g}$$

We need not worry too much at this point about the relative energies of the ligand-field terms associated with any free-ion term, though in passing the relationship with orbital splitting can be noted. So, just as a set of *d*-orbitals splits into t_{2g} (low energy) and e_g (high energy) groups in an octahedral field, so a ^1D term splits into low energy $^1T_{2g}$ and higher energy ^1E_g terms.† Similarly, a set of *f*-orbitals in an octahedral field splits into a low energy single orbital of a_{2u} symmetry (f_{xyz}), an intermediate set of three of t_{2u} symmetry ($f_{x(y^2-z^2)}, f_{y(z^2-x^2)}, f_{z(x^2-y^2)}$) and a high energy set of three of t_{1u} symmetry ($f_{x^3}, f_{y^3}, f_{z^3}$) and these are also the relative energies of ligand-field terms arising from a ^1F term.

We can now start to build a correlation diagram which links the free-ion terms to those of a weak field and, ultimately, what happens as Δ_o becomes larger (multiplicities shown in parentheses):

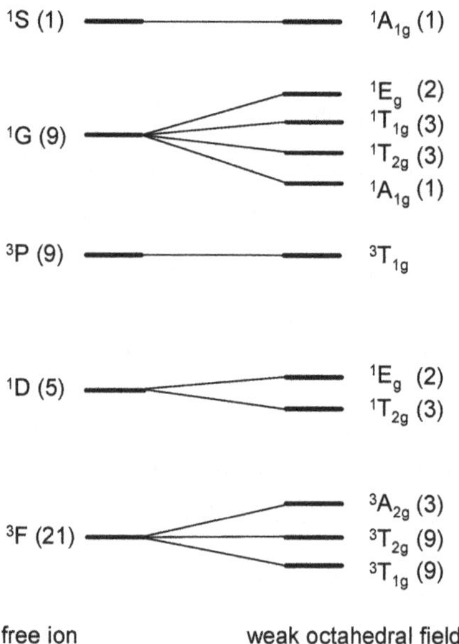

† Note that this is not true for d^8 which also gives rise to a ^1D term. By the hole formalism, it is ^1E_g that is lower in energy than $^1T_{2g}$. Similar comments apply to the ^3F term and d^2 / d^8 systems. This is discussed further in Section 13.4.

As Δ_o increases, so the separation between t_{2g} and e_g levels increases, changing the energy of the microstates associated with the weak field. When Δ_o is large it ultimately determines the energies of the different distributions of electrons between the available *d*-orbitals, significantly so in the $d^4 - d^7$ cases where both high and low-spin alternatives are possible ground states. Using a d^2 configuration by way of illustration, there are three possible configurations for the two electrons across the t_{2g} and e_g orbitals:

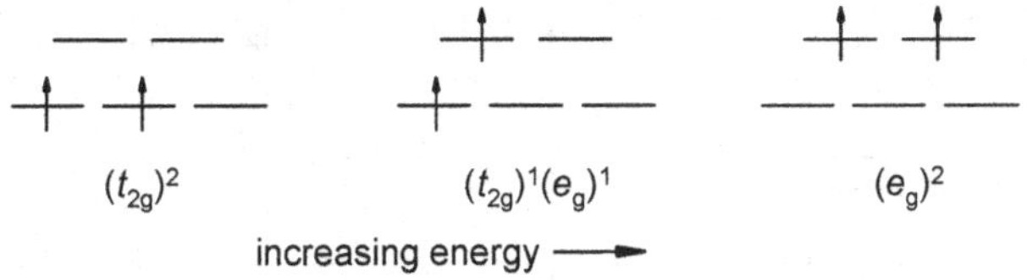

$(t_{2g})^2$ $(t_{2g})^1(e_g)^1$ $(e_g)^2$

increasing energy ⟶

The diagrams show three distinct electronic configurations, whose energy increases from left to right in steps of Δ_o. Each case comprises a number of microstates which can be separated into groups, just as the microstates for the free-ion term were spilt into groups in a weak ligand field. The microstates and their ligand-field term symbols from the weak field will be retained in a strong field but their energies will change, as illustrated by the example at the outset of this section. To understand and rationalise these changes brings us back to the the problem that remained unresolved at the end of the previous section: what are the ligand-field terms associated with $(t_{2g})^2$ and $(e_g)^2$ and more specifically, what the multiplicities of these terms are. To recap, we left the following ambiguities about these ligand-field terms:

$$(t_{2g})^2 = {}^1A_{1g} + {}^1E_g + {}^1T_{1g} + {}^3T_{2g} \text{ or } {}^1A_{1g} + {}^1E_g + {}^3T_{1g} + {}^1T_{2g} \quad \text{(15 microstates)}$$

$$(e_g)^2 = {}^1A_{1g} + {}^3A_{2g} + {}^1E_g \text{ or } {}^3A_{1g} + {}^1A_{2g} + {}^1E_g \quad \text{(6 microstates)}$$

For completeness and in progressing the analysis of the term multiplicities for $(t_{2g})^2$ and $(e_g)^2$, terms for the $(t_{2g})^1(e_g)^1$ configuration also have to be determined. Using the direct product method, these are:

$$(t_{2g})^1(e_g)^1 = T_{2g} \times E_g = T_{1g} + T_{2g}$$

These T_{1g}, T_{2g} terms can be either singlet or triplet multiplicities, while the total degeneracy of the $(t_{2g})^1(e_g)^1$ state is given by combining the degeneracy of its two components individually (eqn. 12.1):

$$(t_{2g})^1 : D_t = (2 \times 3)! \, / \, 1! \times 5! \; = 6$$

$$(e_g)^1 : D_t = (2 \times 2)! \, / \, 1! \times 3! \; = 4$$

The overall degeneracy of the $(t_{2g})^1(e_g)^1$ configuration is therefore $6 \times 4 = 24$ microstates. Along with the 15 microstates for $(t_{2g})^2$ and 6 for $(e_g)^2$ we have accounted for the 45 microstates for d^2. We now need to find 24 microstates which correspond to the ligand-field terms ${}^{1,3}T_{1g}$ + ${}^{1,3}T_{2g}$, for which only one possibility exists:

$$(t_{2g})^1(e_g)^1 = {}^1T_{1g} + {}^1T_{2g} + {}^3T_{1g} + {}^3T_{2g} \quad (3 + 3 + 9 + 9 = 24 \text{ microstates})$$

These T_{1g} / T_{2g} pairs of terms relate to electons distributed between t_{2g} and e_g levels but either in orbitals in different planes e.g. $(d_{xy})^1(d_{z^2})^1$ (lower in energy) or in the same plane e.g. $(d_{xy})^1(d_{x^2-y^2})^1$ (higher in energy). That is, if the electrons are in the same region of space they repel more and are of higher energy.

To return, then, to the issue of multiplicities for ligand-field terms associated with $(t_{2g})^2$ and $(e_g)^2$. To summarise what has been determined so far:

Free-ion term	Ligand-Field term	d^n Config'n	Ligand-Field term
^1S	$^1A_{1g}$	$(t_{2g})^2$	$^1A_{1g} + {}^1E_g + {}^1T_{1g} + {}^3T_{2g}$ *or* $^1A_{1g} + {}^1E_g + {}^3T_{1g} + {}^1T_{2g}$
^3P	$^3T_{1g}$	$(t_{2g})^1(e_g)^1$	$^1T_{1g} + {}^1T_{2g} + {}^3T_{1g} + {}^3T_{2g}$
^1D	$^1E_g + {}^1T_{2g}$	$(e_g)^2$	$^1A_{1g} + {}^3A_{2g} + {}^1E_g$ *or* $^3A_{1g} + {}^1A_{2g} + {}^1E_g$
^3F	$^3A_{2g} + {}^3T_{1g} + {}^3T_{2g}$		
^1G	$^1A_{1g} + {}^1E_g + {}^1T_{1g} + {}^1T_{2g}$		

In a weak ligand field, the ^3F ground term for d^2 splits into $^3A_{2g} + {}^3T_{1g} + {}^3T_{2g}$. $^3A_{2g}$ appears only once among the ligand-field terms for $(t_{2g})^2$, $(e_g)^2$ and $(t_{2g})^1(e_g)^1$, and, since terms in the weak field must correlate with those in the strong field (i.e. must be present in both cases) $^3A_{2g}$ must be present in both cases. On this basis, the ligand-field terms for $(e_g)^2$ must include $^3A_{2g}$ and be :

$$(e_g)^2 = {}^1A_{1g} + {}^3A_{2g} + {}^1E_g$$

We are then left with deciding which of the terms T_{1g} or T_{2g} associated with the $(t_{2g})^2$ configuration is the spin triplet. ^3F includes both these terms ($^3T_{1g} + {}^3T_{2g}$), but we have already established the presence of a $^3T_{2g}$ term among those for the $(t_{2g})^1(e_g)^1$ configuration. Since the weak field $^3T_{2g}$ can only correlate with one such term in the strong field and this has been accounted for means that the term arising from the $(t_{2g})^2$ configuration must be $^3T_{1g}$, that is:

$$(t_{2g})^2 = {}^1A_{1g} + {}^1E_g + {}^3T_{1g} + {}^1T_{2g}$$

These findings can be fully summarized in a complete correlation diagram for d^2 (Fig. 12.2, *overleaf*). To avoid ambiguity in discussing this figure, the two $^3T_{1g}$ terms are distinguished by the symbolism $^3T_{1g}(F)$ and $^3T_{1g}(P)$ to reflect their origin. Furthermore, in linking terms from weak to strong fields an alternative correlation would be that $^3T_{1g}(P)$ becomes the $^3T_{1g}$ of $(t_{2g})^2$, and that $^3T_{1g}(F)$ correlates with $(t_{2g})^1(e_g)^1$. This is not the case because of what is termed the **no crossing rule**. This states that terms of the same symmetry and multiplicity repel one another, so increasing the separation between them. This applies to both ground and excited terms, but luckily only an F ground term (d^2, d^3, d^7, d^8) has an excited term (and hence ligand-field terms) of the same multiplicity.[†]

Correlation diagrams such as Fig. 12.2 can be somewhat daunting and their relationship to actual electron distributions lost, so a synopsis and some examples will help clarify what the diagram is telling us.

[†] This rule also applies to two excited states with the same term symbol, but is of less significance than terms which match the ground term.

Firstly, a summary of what determines the overall energy of the electron distribution:

- for a free ion, the five *d*-orbitals are degenerate; energy is only determined by $e - e$ interactions, with maximum S having lowest energy. This is shown on the extreme left of Fig. 12.2.
- in a ligand field ($\Delta_o > 0$) the degeneracy of the *d*-orbitals is lowered to t_{2g} and e_g.
- for very small Δ_o (a weak field) the energies of $e - e$ interactions are more significant than the small difference in energy between available *d*-orbitals.
- in a weak field the lowest energy term (set of microstates) has maximum multiplicity, as it does for the free ion; this is shown to the right of the free ion in Figure 12.2.

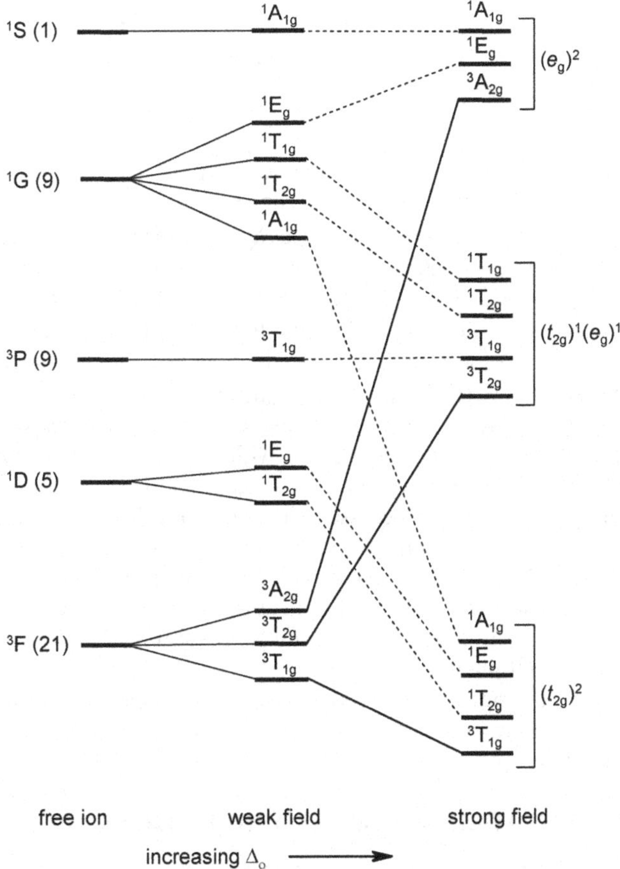

Fig. 12.2 Splitting of the d^2 free-ion terms in an increasing ligand field.

- when Δ_o gets bigger, the energy difference between orbitals starts to outweight the $e - e$ interactions and begins to dominate the total energy of the system.
- when Δ_o is infinitely large, $e - e$ interactions can be ignored and it is more convenient to associate groups of microstates with a $(t_{2g})^x(e_g)^y$ configuration; this is shown on the far right of Figure 12.2.
- in a strong, but not infinitely large, ligand field, the population of the orbital(s) primarily determines the relative energies of different electron distributions, with spin multiplicity playing a secondary role; this is shown to the left of the electronic configurations in Figure 12.2.

A simple example which embodies these ideas is the $^3A_{2g}$ term which arises from the ^3F ground term of the free ion. The two electrons are in the e_g level to give an ^3A term:

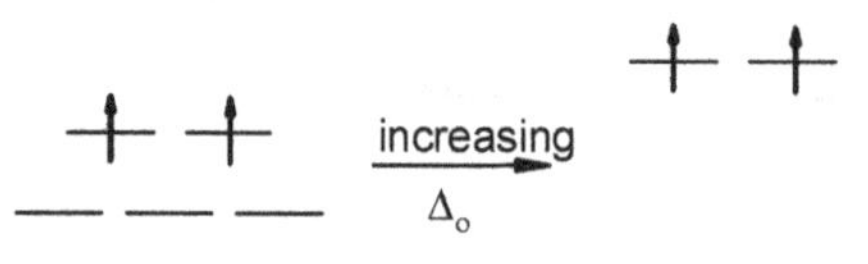

This is a low energy pair of microstates when Δ_o is small because the spins are parallel and inter-electron interactions contribute significantly to the overall energy, but as Δ_o increases these microstates increase rapidly in energy as the influence of the ligands becomes more pronounced (increasing Δ_o) and the energy of the e_g orbitals increases.

What are crucial within Figure 12.2 are the spin-allowed transitions from the ground state, which, irrespective of the magnitude of Δ_o, are $^3T_{2g} \leftarrow {}^3T_{1g}(F)$, $^3T_{1g}(P) \leftarrow {}^3T_{1g}(F)$ and $^3A_{2g} \leftarrow {}^3T_{1g}(F)$.† This is a conclusion we could have arrived at without the full correlation diagram, as these transitions all involve terms arising from splitting of the free-ion ground term along with the only excited free-ion term of triplet multiplicity; all the other terms on the right hand side of Figure 12.2, though completing the picture, are not necessary in predicting the spin-allowed electronic transitions. For all species which have a ground state determined by maximum S i.e. all tetrahedral complexes and all octahedral ones which are not low-spin (d^4 – d^7 only), this will be the case. This simplification, which we discuss more fully in Section 13.4, does not, however, apply to low-spin species where the ground term is not that of highest multiplicity.

Finally, which of the spin-allowed transitions we have identified above are allowed by the symmetry selection rules have yet to be determined and this will be covered in the next chapter.

12.5 SYMMETRY LABELS FOR d^n CONFIGURATIONS – CONCLUSION

Section 12.3 ended with a list of terms associated with any distribution of ten electrons across the five *d*-orbitals (Table 12.6) but with some, but not all, the

---

† The nomenclature for electronic transitions which involve absorption of energy is: higher state ← lower state.

associated spin multiplicities identified. The methodology of Section 12.4 has resolved these issues and Table 12.6 can now be up-dated to incorporate all the spin multiplicites, with the ground term (highest multiplicity) highlighted.

config^n	Term symbols	config^n	Term symbols
$(t_{2g})^1$	$\mathbf{^2T_{2g}}$	$(e_g)^1$	$\mathbf{^2E_g}$
$(t_{2g})^2$	$\mathbf{^3T_{1g}} + {}^1A_{1g} + {}^1E_g + {}^1T_{2g}$	$(e_g)^2$	$\mathbf{^3A_{2g}} + {}^1A_{1g} + {}^1E_g$
$(t_{2g})^3$	$\mathbf{^4A_{2g}} + {}^2E_g + {}^2T_{1g} + {}^2T_{2g}$	$(e_g)^3$	$\mathbf{^2E_g}$
$(t_{2g})^4$	$\mathbf{^3T_{1g}} + {}^1A_{1g} + {}^1E_g + {}^1T_{2g}$	$(e_g)^4$	$\mathbf{^1A_{1g}}$
$(t_{2g})^5$	$\mathbf{^2T_{2g}}$		
$(t_{2g})^6$	$\mathbf{^1A_{1g}}$		

This now allows the term symbols for all the possible d^n configurations to be evaluated by taking direct products of any combination of the configurations in Table 12.6 (*above*). However, we need only focus on direct products between the lowest energy terms for a given pair of configurations as these will give the lowest energy term(s) for the product. Two examples will illustrate this:

$$d^8, (t_{2g})^6(e_g)^2 : \quad {}^1A_{1g} \times {}^3A_{2g} = {}^3A_{2g}$$

$$\text{high-spin } d^4, (t_{2g})^3(e_g)^1 : \quad {}^4A_{2g} \times {}^2E_g = {}^5E_g$$

In both cases the rules of Tables 11.1 and 12.5 have been employed. The maximum spin multiplicities are found but considering S for each term and summing e.g. in the case of $(t_{2g})^3(e_g)^1$, the ${}^4A_{2g}$ and ${}^2E_g$ terms indicate S of $^3/_2$ and $^1/_2$, respectively ($2S + 1 = 4$ or 2), so for the direct product $S = 2$ ($^3/_2 + {}^1/_2$) and $2S + 1 = 5$. Note that this summation only gives the maximum spin; other possible spin multiplicities are dealt with as necessary in the next chapter (*Section 13.6*).

SAQ 12.6 : What are the symmetry labels for the ground terms of both high- and low-spin d^7 configurations in a octahedral ligand field ?

This process leads to the rather daunting looking Table 12.7 (*overleaf*), though it can be broken down and be replaced by a pair of simple diagrams which will be presented in Section 13.4. The table of ground terms for all d^n configurations allows us to evaluate and understand the possible spin-allowed transitions for the general case of an octahedral transition metal complex in an intermediate ligand field, that is one in which the magnitude of Δ_o is similar to the $e - e$ repulsion energies.

12.6 SUMMARY

- the degeneracy of a configuration (number of microstates) is given by:

  $D_t = N! / (N_e)!(N_h)!$

- a *term* is a label which describes a collection of microstates.

- for a free ion, the ground state term has maximum S and maximum L.
- terms are split in a ligand field, just as atomic orbitals are.
- in a weak ligand field (small Δ_o) the multiplicity of the ground term is the same as that for the free ion, though this is not necessarily true in a strong ligand field.
- the energies of the ligand-field terms change with increasing Δ_o.
- the *no crossing rule* states that terms of the same symmetry and multiplicity repel one another, so increasing the separation between them as Δ_o increases.
- the ground terms for any $(t_{2g})^x(e_g)^y$ configuration can be determined by the direct product of the ground terms for the individual $(t_{2g})^x$ and $(e_g)^y$ configurations.

Table 12.7 Ground terms for $(t_{2g})^x(e_g)^y$ configurations

	Configuration	Ground term	Configuration	
d^1	$(t_{2g})^1$	$^2T_{2g}$	$(t_{2g})^5(e_g)^4$	d^9
	$(e_g)^1$	^2E_g	$(t_{2g})^6(e_g)^3$	
d^2	$(t_{2g})^2$	$^3T_{1g}$	$(t_{2g})^4(e_g)^4$	d^8
	$(t_{2g})^1(e_g)^1$	$^3T_{1g} + {}^3T_{2g}$	$(t_{2g})^5(e_g)^3$	
	$(e_g)^2$	$^3A_{2g}$	$(t_{2g})^6(e_g)^2$	
d^3	$(t_{2g})^3$	$^4A_{2g}$	$(t_{2g})^3(e_g)^4$	d^7
	$(t_{2g})^2(e_g)^1$	$^4T_{1g} + {}^4T_{2g}$	$(t_{2g})^4(e_g)^3$	
	$(t_{2g})^1(e_g)^2$	$^4T_{1g}$	$(t_{2g})^5(e_g)^2$	
	$(e_g)^3$	^2E_g	$(t_{2g})^6(e_g)^1$	
d^4	$(t_{2g})^4$	$^3T_{1g}$	$(t_{2g})^2(e_g)^4$	d^6
	$(t_{2g})^3(e_g)^1$	^5E_g	$(t_{2g})^3(e_g)^3$	
	$(t_{2g})^2(e_g)^2$	$^5T_{2g}$	$(t_{2g})^4(e_g)^2$	
	$(t_{2g})^1(e_g)^3$	$^3T_{1g} + {}^3T_{2g}$	$(t_{2g})^5(e_g)^1$	
	$(e_g)^4$	$^1A_{1g}$	$(t_{2g})^6$	
d^5	$(t_{2g})^5$	$^2T_{2g}$		
	$(t_{2g})^1(e_g)^4$	$^2T_{2g}$		
	$(t_{2g})^4(e_g)^1$	$^4T_{1g} + {}^4T_{2g}$		
	$(t_{2g})^2(e_g)^3$	$^4T_{1g} + {}^4T_{2g}$		
	$(t_{2g})^3(e_g)^2$	$^6A_{1g}$		

PROBLEMS

Answers to all problems marked with * are given in Appendix 4.

1*. What is the degeneracy of an $(s)^1(p)^1$ configuration? Determine the terms associated with this configuration and identify which of these is the ground term ?

2. What is the ground term for a d^5 free-ion ?

3. Use eqns. 12.2 – 12.6 to determine the symmetry labels for the *d*-orbitals under D_{4h} symmetry. Check your answer with the character table in Appendix 5.

4*. Confirm, using the appropriate direct product and combination of spin multiplicities, that the ground terms for $(t_{2g})^4(e_g)^1$ are $^4T_{1g}$ + $^4T_{2g}$ and for $(t_{2g})^3(e_g)^2$ is $^6A_{1g}$.

5*. The Pr^{3+} ion has $[Xe](4f)^2$ configuration. What is the ground term for this ion and how is this term split in an octahedral field (use Eqn. 12.2 – 12.6) ?

13
d-d SPECTRA

Literally the most visible distinction between compounds of the *d*-block elements and their main group counterparts is the predominance of colour among the former. Colour - arising from electronic transitions which occur in the visible region of the electromagnetic spectrum - is largely associated with *d-d* transitions, that is between the energy levels (terms) discussed in Chapter 12, though other types of transition e.g. charge transfer are also important. At its simplest, the magenta colour of $[Ti(H_2O)_6]^{3+}$ (d^1) arises from the single electron $e_g \leftarrow t_{2g}$ transition at *ca.* 20,000 cm^{-1} but spectra from other d^n configurations are generally less easy to rationalise. For example, we might expect a d^2 ion such as green $[V(H_2O)_6]^{3+}$ to have two spectral bands corresponding to $e_g \leftarrow t_{2g}$ transitions of either one or both electrons. In fact, two bands are seen in the visible spectrum (17,800, 25,700 cm^{-1}) but not at energies Δ_o and $2\Delta_o$ as might be anticipated, along with other bands of much weaker intensities.

This chapter will use the results of Chapter 12 concerning the terms associated with the d^n configurations to attempt to rationalise these spectra, as far as is reasonable in a text concerned primarily with applications of group theory.

13.1 THE BEER-LAMBERT LAW

The amount of incident light (I_0) transmitted (I_T) after passing through a sample is proportional to both the path length (l) and the concentration (C) of that sample. This is more commonly stated in terms of the absorbance of the sample, in which

$$\text{Absorbance, } A = \log_{10}[(I_0) / (I_T)]$$

$$A \propto l \times C$$

$$A = \varepsilon \times l \times C \qquad \text{(Eqn. 13.1)}$$

Equation 13.1 is the **Beer-Lambert law** and the constant of proportionality ε is the **molar absorption coefficient**. ε has units of $dm^3mol^{-1}cm^{-1}$ and is measured at the absorbance (band) maximum (λ_{max}) for a 1 M solution of path length 1 cm. ε values allow comparison of the probability of a given electronic transition taking place: the more intense the absorption the more likely the transition. Transition probability in turn relates to whether or not a transition is "allowed" or occurs after violation of one of the selection rules, which we will now consider in more depth.

13.2 SELECTION RULES AND VIBRONIC COUPLING

The symmetry selection rule which governs electron transitions was discussed in Section 11.2 and requires a non-zero transition integral for the transition to be allowed:

$$\int \psi_i \mu \psi_f \, d\tau \neq 0 \quad \text{i.e } \psi_i \times \begin{pmatrix} \mu_x \\ \mu_y \\ \mu_z \end{pmatrix} \times \psi_f = \psi_i \times \begin{pmatrix} \mathbf{T_x} \\ \mathbf{T_y} \\ \mathbf{T_z} \end{pmatrix} \times \psi_f \neq 0$$

Since the symmetry of $\mathbf{T_x}$, $\mathbf{T_y}$, $\mathbf{T_z}$ under O_h symmetry is T_{1u}, this becomes:

$$\int \psi_i \mu \psi_f \, d\tau = \psi_i \times T_{1u} \times \psi_f$$

We know that ψ_i and ψ_f both have *g* symmetry since this is the case for all terms arising from a d^n configuration. Thus, considering only the *u* / *g* aspects of the integral:

$$\psi_i \times T_{1u} \times \psi_f = g \times u \times g = u$$

For the integral to be non-zero it must contain the totally symmetric representation, which in the O_h point group is A_{1g}; this is impossible if the product $\psi_i \times T_{1u} \times \psi_f$ has *u* symmetry.

The same analysis also rules out $s \rightarrow s$ and $p \rightarrow p$ transitions. Remembering that *s*-orbitals are always *g* and *p*-orbitals *u* in point groups with a centre of inversion, and that μ has the same symmetry as $\mathbf{T_x}$, $\mathbf{T_y}$, $\mathbf{T_z}$, which in turn have the same symmetry as the *p*-orbitals (*u*), irrespective of the point group we have:

$$s \rightarrow s: \quad \psi_i \times \mu \times \psi_f = g \times u \times g = u$$

$$p \rightarrow p: \quad \psi_i \times \mu \times \psi_f = u \times u \times u = u$$

From this we have the **parity selection rule**:

- a transition must involve a change of parity; $g \rightarrow g$ and $u \rightarrow u$ transitions are parity forbidden.

This rule is, in fact, contained within the more general **Laporte rule**[†] which requires a change in orbital angular momentum for a transition between orbitals to be allowed.

- an allowed transition must have $\Delta l = \pm 1$.

[†] O Laporte and W F Meggers, *J. Opt. Soc. Am.*, **11**, 459 (1925). Note that it is common to find the Laporte and parity rules treated as one and the same.

This means that:

- $s \rightarrow s$, $p \rightarrow p$ and $d \rightarrow d$ are forbidden ($\Delta l = 0$; no change in parity).
- transitions such as $s \rightarrow p$ and $p \rightarrow d$ are allowed ($\Delta l = \pm 1$), as they are on parity grounds.

$d \rightarrow d$ transitions for octahedral species are forbidden both on parity grounds and the lack of change in orbital angular momentum ($\Delta l = 0$), though clearly the colour inherent in most octahedral transition metal complexes suggests *d-d* transitions do take place and both rules are violated, but how does this arise? Simplistically, this is because certain vibrational modes of the molecule remove the inversion symmetry at the heart of the parity issue and allow the transition to take place, a process termed **vibronic coupling**. We can rationalise this more formally in the following way.

Electronic transitions between energy levels are not totally independent of the vibrational motions within the molecule, even though this is generally ignored in an analysis such as we have been developing. The integral of eqn. 11.1 is better written as:

$$\int \psi_i \mu \psi_f \, d\tau \quad = \quad \int (\psi_e \psi_v)_i \mu (\psi_e \psi_v)_f \, d\tau \qquad \text{(eqn 13.2)}$$

The ground and excited states have contributions from both the electronic and vibrational wavefunctions. That is, the symmetry of the ground (and excited) state is the direct product $\psi_e \times \psi_v$; this coupling of vibrational and electronic behaviour gives rise to the term "vibronic". The following points, which have been made earlier, become important:

- the direct product of a non-degenerate irreducible representation and itself is always the totally symmetric representation (*Section 11.2*).
- the direct product of a degenerate irreducible representation and itself is a reducible representation (*Section 11.4*) and includes the totally symmetric representation (Table 11.1).
- the ground vibrational state $(\psi_v)_i$ is always totally symmetric (*Section 11.5*).

To these we can add the rather obvious fact that:

- any binary direct product involving the totally symmetric representation leaves the second component unchanged.

SAQ 13.1 : Show that $A_{1g} \times T_{1u} = T_{1u}$ under O_h symmetry by multiplying the characters of the two irreducible representations.

Answers to all SAQs are given in Appendix 3.

We can now start to simplify eqn. 13.2:

$$\int (\psi_e \psi_v)_i \mu (\psi_e \psi_v)_f \, d\tau \quad = \quad \int (\psi_e)_i \mu (\psi_e)_f (\psi_v)_f \, d\tau$$

since $(\psi_v)_i$ is the totally symmetric representation (A_{1g} under O_h symmetry).

We know that for the O_h point group $(\psi_e)_i \times \mu \times (\psi_e)_f$ has a *u* symmetry label ($g \times u \times g$), and this requires that $(\psi_v)_f$ must also be *u* to make the complete direct product *g*. Furthermore, to make the direct product either be, or contain, the totally symmetric representation then $(\psi_e)_i \times \mu \times (\psi_e)_f$ must contain (or **span**) the same symmetry labels as $(\psi_v)_f$. An example will make this clearer.

For a d^1 system, the spin-allowed (but Laporte-forbidden) transition is $^2E_g \leftarrow {}^2T_{2g}$. Since μ has T_{1u} symmetry in an octahedral field:

$$\int (\psi_e)_i\mu(\psi_e)_f(\psi_v)_f \, d\tau = (T_{2g} \times T_{1u} \times E_g) \times (\psi_v)_f$$

Using the rules of Table 11.1 and breaking down this multiple direct product into a series of binary products:

$$(T_{2g} \times T_{1u} \times E_g) = (T_{2g} \times T_{1u}) \times E_g = (A_{2u} + E_u + T_{1u} + T_{2u}) \times E_g$$

$$\begin{aligned}(A_{2u} + E_u + T_{1u} + T_{2u}) \times E_g &= E_u + A_{1u} + A_{2u} + E_u + T_{1u} + T_{2u} + T_{1u} + T_{2u} \\ &= A_{1u} + A_{2u} + 2E_u + 2T_{1u} + 2T_{2u}\end{aligned}$$

Now the vibrational modes of an octahedral ML_6 species are (*Chapter 4, question 4*):

$$\Gamma_{vib} = A_{1g} + E_g + 2T_{1u} + T_{2g} + T_{2u}$$

and these are what we mean by $(\psi_v)_f$ in eqn. 13.2. So:

$$(T_{2g} \times T_{1u} \times E_g) \times (\psi_v)_f = (A_{1u} + A_{2u} + 2E_u + 2T_{1u} + 2T_{2u}) \times (A_{1g} + E_g + 2T_{1u} + T_{2g} + T_{2u})$$

There is no need to evaluate this daunting series of direct products, as we know that A_{1g}, the totally symmetric representation that is required to make the transition integral non-zero, is (or is contained in) the product of two identical representations; we only need to look for matches in the two halves the multiplication to achieve this outcome. Both T_{1u} and T_{2u} statisfy this criterion and the transition becomes vibronically-allowed. Moreover, when terms in $(\psi_e)_i \times \mu \times (\psi_e)_f$ which span Γ_{vib} occur more than once, then vibronic coupling will split the transition by an amount similar to that of the energy of a metal-ligand bond vibration (*ca.* 200 cm^{-1}), which, in part, accounts for the broad nature of most *d-d* spectral linewidths.

> *SAQ 13.2 : A low-spin* Co^{3+} *complex has* $^1A_{1g}$ *ground state and excited states* $^1T_{1g}$ *and* $^1T_{2g}$*. Does vibronic coupling allow these transitions to occur ?*

Removing the centre of symmetry in an octahedral complex also allows a mixing of *p*- and *d*-orbitals (which is forbidden under strict O_h symmetry; they have different symmetry labels). This means that as well as breaking the strict parity condition ($g \rightarrow g$), vibronic coupling also means that $\Delta l \neq 0$ as the orbitals involved in the transition are not pure *d* in nature.

The concept of parity has, or course, no significance in point groups which lack a centre of inversion e.g. T_d, though $d \rightarrow d$ transitions are still forbidden as $\Delta l = 0$. However, mixing of *p*- and *d*-orbitals is symmetry-allowed under T_d symmetry (*p*-orbitals have the same t_2 symmetry as d_{xy}, d_{xz}, d_{yz}) and this provides a way of relaxing the Laporte rule i.e. transitions are no longer between pure *d*-orbitals. As a result of no parity restrictions, the colour of tetrahedral complexes is generally more intense (ε larger by one or two orders of magnitude) than that of octahedral complexes as the $d \rightarrow d$ transition is Laporte-allowed.†

13.3 THE SPIN SELECTION RULE

A simple extension of the analysis which gives rise to the Laporte rule also justifies the origin of the spin selection rule.

- for a spin-allowed transition $\Delta S = 0$, i.e. there is no change in electron spin

Since the energy of an electronic state has contributions from both spin (S) and orbital (L) considerations, eqn. 13.2 should be further expanded as:

$$\int(\psi_e)_i\mu(\psi_e)_f(\psi_v)_f\, d\tau \;=\; \int(\psi_S)_i(\psi_L)_i\mu(\psi_S)_f(\psi_L)_f(\psi_v)_f\, d\tau \quad \text{(Eqn. 13.3)}$$

where ψ_e has been replaced by $(\psi_S)(\psi_L)$ in both ground and excited states. Since we know that a symmetry-allowed transition has $(\psi_L)_i \times \mu \times (\psi_L)_f \times (\psi_v)_f$ which spans the totally symmetric reducible representation *before* considering the inclusion of spin i.e. the $(\psi_S)_i \times (\psi_S)_f$ component, this latter product must itself be totally symmetric to retain the overall total symmetry of the integral. This requires that $(\psi_S)_i = (\psi_S)_f$ and this will only be true if both ground and excited states have the same spin, S.

However, the $[Mn(H_2O)_6]^{2+}$ ion is very pale pink, despite the fact that it is high-spin d^5 and the $^6A_{1g}$ ground term is not split by a ligand field nor are there any excited states of sextet multiplicity (Table 12.3), so it is evident that the spin selection rule is being broken. On the other hand, the lack of intensity arising from whatever *d-d* transition(s) give rise to the pink colouration suggests that the rule is hard to break. Violation of the spin selection rule occurs because the spin and orbital motion of the electrons cannot be segregated, they couple in much the way vibronic coupling causes relaxation of the Laporte rule. Spin-orbit coupling, which has been touched on previously at the end of Section 12.1, is weak for first row transition elements like manganese which conform to the Russell-Saunders (L, S) scheme. For second- and third-row transition elements the extent of spin-orbit coupling increases, resulting in more significant relaxation of the spin selection rule. In fact, for these elements the Russell-Saunders coupling scheme becomes increasingly inappriopriate: spin-orbit coupling is dominant and the outcome is perturbed by the ligand-field, whereas in the Russell-Saunders scheme the effect of the ligand field is dominant and L, S coupling has a relatively minor effect. For heavier elements it becomes more appropriate to couple the spin and orbital angular momentum of individual electrons (j - j coupling), then sum these individual j values to give J.

Table 13.1 (*overleaf*) summarises the relative intensities of bands of different origin in the light of the various selection rules. Charge transfer bands can arise from either donation of electrons from a filled metal *d*-orbital to an empty (or partially empty) ligand orbital (**metal-to-ligand** or **MLCT**), or from a filled ligand orbital to an empty (or partially empty) metal *d*-orbital (**ligand-to-metal** or **LMCT**). MLCT is common in

† This is not true if the colour of an octahedral complex arises from something other than a *d-d* transition. Charge transfer bands, such as give rise to the intense purple colour of the permanganate ion, $[MnO_4]^-$ (d^0), are Laporte- and parity-allowed.

complexes with π-acceptor ligands such as CO, CN^-, where the empty π* orbitals on the ligand accept the transferred charge.

Table 13.1 Band intensities and selection rules [a]

Origin	Laporte	Parity	Spin	ε [b]
d-d (O_h)	x	x	x	0.1
d-d (T_d)	x	n/a	x	1
d-d (O_h)	x	x	✓	10
d-d (T_d)	x	n/a	✓	100
Charge transfer	✓	✓	✓	1000 +

[a] x = forbidden, ✓ = allowed
[b] approximate magnitude, $dm^3mol^{-1}cm^{-1}$

Similarly, the intense purple colour of $[MnO_4]^-$ arises from a LMCT to the empty *d*-orbitals on manganese(VII) (d^0). Since the ligand orbitals involved are not *d*-orbitals the Laporte rule is never violated, nor is the spin selection rule as the electron moves to an empty site and retains its original spin. Charge transfer bands are thus almost always much more intense than *d-d* bands.

13.4 *d-d* SPECTRA – HIGH-SPIN OCTAHEDRAL COMPLEXES

Knowing both the terms and their multiplicities for any d^n configuration (Table 12.7) and the impact of selection rules on transition intensities, we can now begin to evaluate the observed electronic spectra for various transition metal ions, begining with octahedral complexes of high-spin as these have ground terms of maximium multiplicity.

For all the high-spin arrangements, we can begin to group d^n configurations together, starting with d^1, d^9, d^4 and d^6 all of which have a *D* free-ion ground term (Table 12.3). These split into T_{2g} and E_g terms in a ligand field, with the energies reversed for d^1 and d^9. That is, while the t_{2g} orbitals are lower in energy in both cases, the hole formalism reverses the energies of the ligand-field terms:

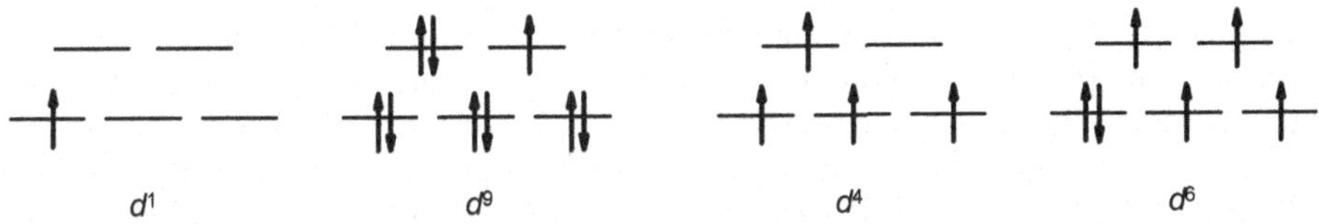

For d^9, the ground state $(t_{2g})^6(e_g)^3$ has a hole in the e_g level, which can be in either d_{z^2} or $d_{x^2-y^2}$ so is an E term i.e. ^2E_g. Similarly, the excited $(t_{2g})^5(e_g)^4$ configuration corresponds to a hole in the t_{2g} level (equivalent to the single d^1 electron) which can occur in any of the three *d*-orbitals and is a T term ($^2T_{2g}$). By analogy, d^6 follows d^1 in having one electron more than a half-filled orbital set and so is $^5T_{2g}$ in its ground state, while d^4 has a hole in a half-filled shell (^5E_g) and parallels d^9; the ground terms for d^4 and d^6 are reversed, namely ^5E_g and $^5T_{2g}$, respectively. These outcomes can be pictorially represented by Fig. 13.1, in which the appropriate multiplicities need to be added in accordance with the relevant d^n case:

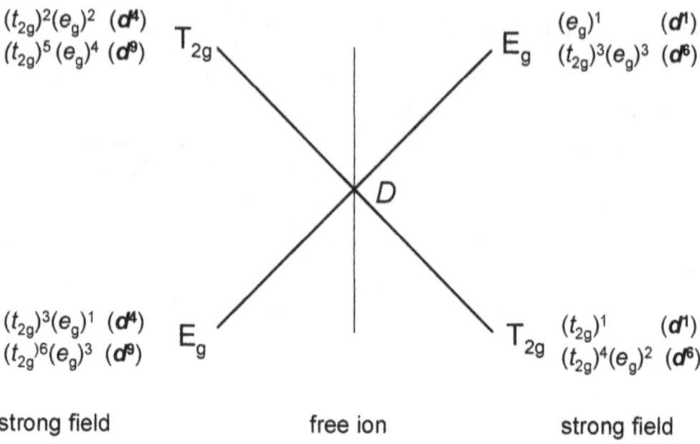

Fig. 13.1 Splitting of a *D* ground term for d^1, d^4, d^6 and d^9 octahedral complexes.

In all the above cases we would expect one spin-allowed transition between ground and excited state. For d^1 this is $^2E_g \leftarrow {}^2T_{2g}$, while for d^9 it is $^2T_{2g} \leftarrow {}^2E_g$.

SAQ 13.3 : What is the spin-allowed transition for a high-spin Co^{3+} complex ?

By the same hole formalism d^2, d^3, d^7 and d^8 configurations, which all have an *F* free-ion ground term can be grouped. Thus, while the d^2 terms of maximum multiplicity are, in ascending order of energy, $^3T_{1g}$, $^3T_{2g}$, $^3A_{2g}$, then for d^8 the order is reversed: $^3A_{2g}$ lowest, then $^3T_{2g}$ and finally $^3T_{1g}$. In fact, the complete correlation diagram for d^8 is arrived at by inverting Fig. 12.2, as shown in Fig. 13.2 (*overleaf*). Note that some of the lines linking $^3T_{1g}$ terms have changed to obey the no crossing rule.

The relationships within the d^2, d^3, d^7 and d^8 family, then, are:

$$d^2 \equiv d^7 \quad \text{and} \quad d^3 \equiv d^8$$

$$d^2, d^7 \text{ are the reverse of } d^3, d^8$$

By analogy, Fig. 13.3 (P. 154) summarises and simplifies the correlation diagram for d^2, d^3, d^7 and d^8, all of which have *F* ground terms and an excited *P* term of the same multiplicity. The curvature in the lines associated with the two T_{1g} terms is a result of the no crossing rule, whereby terms of the same symmetry and multiplicity repel each other and don't cross. On the basis of Fig.13.3 we would predict three possible spin-allowed transitions for these four d^n configurations.

SAQ 13.4 : What are the spin-allowed transitions for $[V(H_2O)_6]^{3+}$?

In practice one of these transitions is extremely weak and at high energy (*ca.* 32,000 cm^{-1}) as it involves the simultaneous excitation of two electrons e.g. for d^2 the $^3A_{2g} \leftarrow {}^3T_{1g}(F)$ transition corresponds to $(e_g)^2 \leftarrow (t_{2g})^2$.

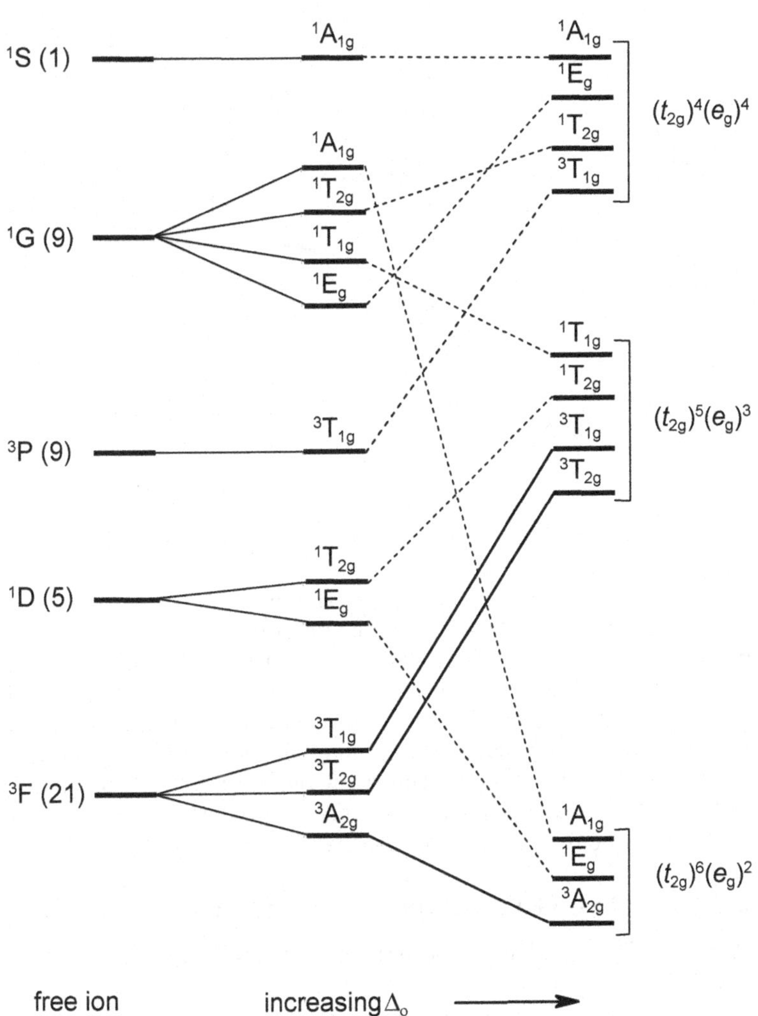

Fig. 13.2 Splitting of the d^8 free-ion terms in an increasing ligand field.

Figures 13.1 and 13.3 are often referred to as **Orgel diagrams**. They offer a straightforward but simplified analysis of *d-d* spectra but fail to explain spectral fine detail by focusing solely on spin-allowed transitions between terms of the same multiplicity. For example, the visible spectrum of the $[V(H_2O)_6]^{3+}$ also includes several extremely weak bands in the 20,000 - 30,000cm^{-1} region, which, based on their intensity ($\varepsilon \approx 0.1$ dm^{3}mol^{-1}cm^{-1}), must arise from spin-forbidden transitions to excited singlet terms.

These excited terms of differing multiplicity to the ground term, absent in Orgel diagrams, are evident in the full correlation diagrams (Fig. 12.2, Fig. 13.2), which,

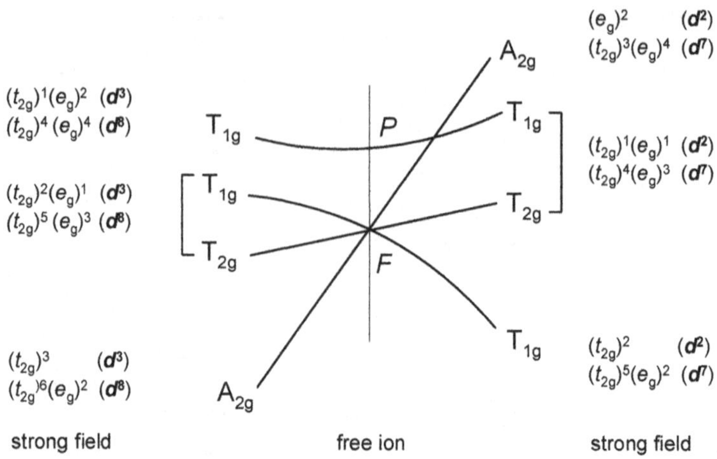

Fig. 13.3 Splitting of a *F* ground term for d^2, d^3, d^7 and d^8 octahedral complexes.

along with other diagrams of this type, are available in a modified and more quantitative form known as **Tanabe-Sugano diagrams.**†

Finally, we have already noted (*Section 13.3*) that high-spin d^5 complexes are virtually colourless since there can be no spin-allowed transitions from the ground term: promotion of an electron must lead to a change in multiplicity. This is also consistent with the fact that there is no splitting of the $^6A_{1g}$ ground term and no excited state sextet terms. The weak transitions giving rise to the pale pink colour of $[Mn(H_2O)_6]^{2+}$ have been assigned to, *inter alia*, spin-forbidden $^4T_{1g}(G) \leftarrow {}^6A_{1g}$ and $^4T_{2g} \leftarrow {}^6A_{1g}$ transitions on the basis of Tanabe-Sugano diagrams.

13.5 *d-d* SPECTRA – TETRAHEDRAL COMPLEXES

Using eqns 12.2 – 12.6 we can determine how any of the free-ion terms split in a tetrahedral field:

T_d	E	$8C_3$	$3C_2$	$6S_4$	$6\sigma_d$	
Γ_S	1	1	1	1	1	$= A_1$
Γ_P	3	0	-1	1	-1	$= T_1$
Γ_D	5	−1	1	-1	1	$= T_2 + E$
Γ_F	7	1	-1	-1	-1	$= A_2 + T_1 + T_2$

The outcome is identical to the case for an octahedral field, though the subscripts *g*, *u* are missing as these do not occur in a point group lacking *i* symmetry. However, the relative energies of the terms have now changed. In a tetrahedral field, the two *d*-orbitals which make up the *e* set (d_{z^2}, $d_{x^2-y^2}$) are lower in energy than the t_{2g} set of three *d*-orbitals. It follows then that for a d^1 configuration the E term is the ground state

† Y Tanabe and S Sugano, *J. Phys. Soc., Jpn*, 1954, **9**, 753; *ibid.*, 766. These are also commonly reproduced in most undergraduate texts on inorganic chemistry.

under T_d symmetry. Conversely, for d^9, the ground term is T_2, the reverse of octahedral d^9. In summary:

$$\text{tetrahedral } d^n \equiv \text{octahedral } d^{(10-n)}$$

$$\text{tetrahedral } d^{(10-n)} \equiv \text{octahedral } d^n$$

The Orgel diagrams (Fig. 13.1, 13.3) can be generalised to cover both octahedral and tetrahedral cases, simultaneously showing these relationships. In both Figures 13.4 and 13.5, the appropriate multiplicities need to be added, and, for octahedral complexes, *g* subscripts also added to the term symbols.

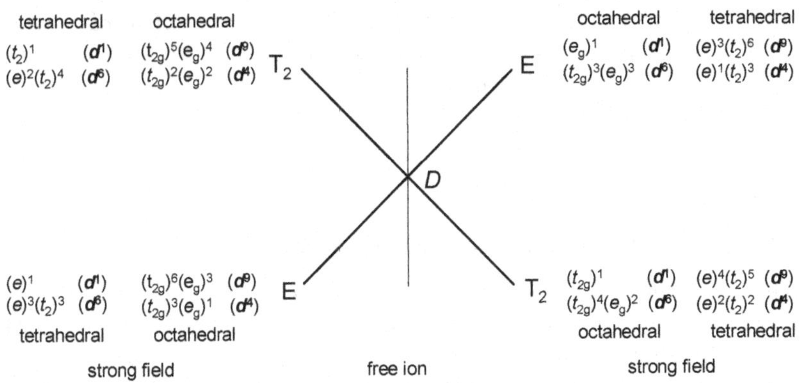

Fig. 13.4 Splitting of a *D* ground term for octahedral and tetrahedral d^1, d^4, d^6 and d^9; appropriate multiplicities and *g* subscripts (for O_h species) need to be added to the term symbols.

SAQ 13.5 : What spin-allowed transitions would be predicted for $[FeCl_4]^{2-}$?

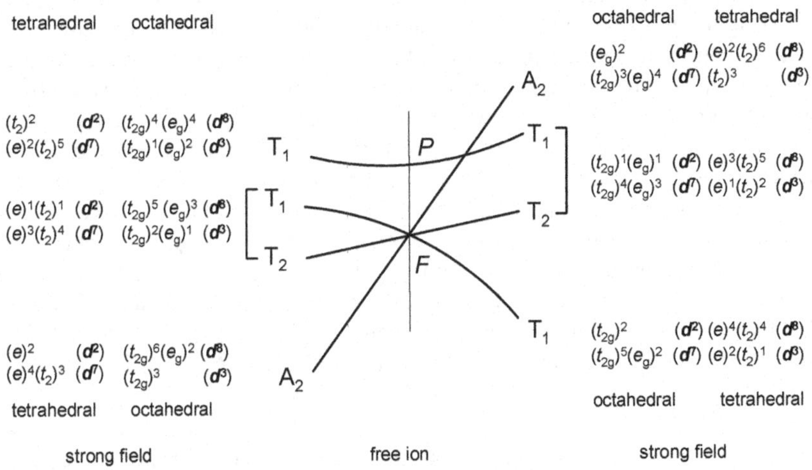

Fig. 13.5 Splitting of a *F* ground term for octahedral and tetrahedral d^2, d^3, d^7 and d^8; appropriate multiplicities and *g* subscripts (for O_h species) need to be added to the term symbols.

SAQ 13.6 : What spin-allowed transitions would be predicted for $[NiCl_4]^{2-}$?

It has already been noted that the spin-allowed *d-d* transitions for octahedral complexes as predicted by Figs. 13.4, 13.5 are only allowed through relaxation of the Laporte / parity rule by vibronic coupling. For tetrahedral complexes it might be assumed that all the *d-d* transitions predicted by these diagrams would be allowed, since the parity rule does not apply to non-centrosymmetric systems. This is not, however, the case, and the zero or non-zero nature of the transition integral still has to be evaluated, as the following *SAQ* illustrates.

SAQ 13.7 : A tetrahedral d^2 ion such as $[VCl_4]^-$ has three plausible spin-allowed transitions associated with promotion of an electron from $(e)^2$ to either $(e)^1(t_2)^1$ or $(t_2)^2$ excited states.
Use Fig. 13.5 to identify these transitions, then evaluate the three appropriate transition integrals to see which of them is symmetry-allowed.

13.6 *d-d* SPECTRA – LOW-SPIN COMPLEXES

Before concluding this discussion it is pertinent to comment very briefly on the electronic spectra of low-spin complexes, which can arise for d^4 - d^7 octahedral complexes in a strong ligand field.

One significant practical issue with low-spin octahedral complexes is their relative scarcity. These only occur with strong-field ligands (i.e. at the high end of the spectrochemical series), such as CN^-. Unfortunately, ligands such as this have empty π^* orbitals available for MLCT bands which, due to their intensity relative to *d-d* bands, often obscure the latter in the spectrum.

From a group theory perspective we can begin to appreciate the complexities that arise for low-spin complexes, embodied in the relevant Tanabe-Sugano diagrams, by considering the possible spin-allowed transitions for such species. The entries for the various d^n configurations given in Table 12.7 are still applicable, but the ground term is no longer the configuration of maximum multiplicity. For example, while high-spin d^5 has a $(t_{2g})^3(e_g)^2$ configuration and $^6A_{1g}$ ground term, low-spin d^5 is $(t_{2g})^5$ with $^2T_{2g}$ lowest in energy. For spin-allowed transitions to be evaluated, excited-state terms with the same multiplicity as the ground state need to be identified and this we have not fully done to date. Again, using low-spin d^5 as an example, we would expect the most likely *d-d* transition to be $(t_{2g})^4(e_g)^1 \leftarrow (t_{2g})^5$, but so far Table 12.6 only identifies the $^4T_{1g}$ and $^4T_{2g}$ terms associated with $(t_{2g})^4(e_g)^1$ as these are the terms of highest multiplicity. What we need to evaluate are other terms arising from $(t_{2g})^4(e_g)^1$ but which have doublet multiplicity to go with the $^2T_{2g}$ ground state of $(t_{2g})^5$. Of course, this is done by the familiar direct product method (Table 12.6, p. 143):

$$(t_{2g})^4 \times (e_g)^1 = (^3T_{1g} + {}^1A_{1g} + {}^1E_g + {}^1T_{2g}) \times {}^2E_g$$

To begin with we will ignore the spin multiplicities and concentrate on the symmetry labels alone. The above direct product can be broken down into the following binary parts:

$$\begin{aligned}T_{1g} \times E_g &= T_{1g} + T_{2g}\\ A_{1g} \times E_g &= E_g\\ E_g \times E_g &= A_{1g} + A_{2g} + E_g\\ T_{2g} \times E_g &= T_{1g} + T_{2g}\end{aligned}$$

The spin multiplicities associated with these terms combine vectorially to give $S_1 + S_2$, $S_1 + S_2 - 1, \ldots\ |S_1 - S_2|$, as summarised in the following Table:

Table 13.2 Combinations of term multiplicities

Term 1 (S_1)	Term 2 (S_2)	Term 1 × Term 2
singlet	singlet	doublet
singlet	doublet	doublet
singlet	triplet	triplet
singlet	quartet	quartet
doublet	doublet	triplet + singlet
doublet	triplet	quartet + doublet
doublet	quartet	quintet + triplet
triplet	triplet	quintet + triplet + singlet
triplet	quartet	sextet + quartet + doublet

To put this table into perspective, consider the final entry in which $S_1 = 1$ (triplet) and $S_2 = {}^3/_2$ (quartet) terms are combined. For S_1, $M_S = 1, 0$ or -1, while for S_2, $M_S = {}^3/_2, {}^1/_2, -{}^1/_2$ or $-{}^3/_2$. Combining these individual M_S values gives the following summations:

$$\begin{aligned}S_2 + 1 &: \ {}^5/_2, {}^3/_2, {}^1/_2, -{}^1/_2\\ S_2 + 0 &: \ {}^3/_2, {}^1/_2, -{}^1/_2, -{}^3/_2\\ S_2 + (-1) &: \ {}^1/_2, -{}^1/_2, -{}^3/_2, -{}^5/_2\end{aligned}$$

These combinations can be broken down and grouped together in the same way that term symbols for microstates are arrived at (*Appendix 2*). Starting with the highest $M_s = {}^5/_2$, this is part of a sextet multiplicity in which $S = {}^5/_2$ ($S_1 + S_2$), and which also includes $M_s = {}^3/_2, {}^1/_2, -{}^1/_2, -{}^3/_2$ and $-{}^5/_2$. Removing these from the group above leaves the highest remaining $M_s = {}^3/_2$ ($S_1 + S_2 - 1$), which is part of a quartet multiplicity ($S = {}^3/_2$), which also includes $M_s = {}^1/_2, -{}^1/_2, -{}^3/_2$. This leaves only two M_s values (${}^1/_2, -{}^1/_2$) which imply a doublet ($S = {}^1/_2 = |S_1 - S_2|$).

The binary direct products for $(t_{2g})^4 \times (e_g)^1$ therefore become:

$$\begin{aligned}{}^3T_{1g} \times {}^2E_g &= {}^4T_{1g} + {}^4T_{2g} + {}^2T_{1g} + {}^2T_{2g}\\ {}^1A_{1g} \times {}^2E_g &= {}^2E_g\\ {}^1E_g \times {}^2E_g &= {}^2A_{1g} + {}^2A_{2g} + {}^2E_g\\ {}^1T_{2g} \times {}^2E_g &= {}^2T_{1g} + {}^2T_{2g}\end{aligned}$$

So overall:

$$\begin{aligned}(t_{2g})^4 \times (e_g)^1 &= {}^4T_{1g} + {}^4T_{2g} + {}^2T_{1g} + {}^2T_{2g} + {}^2E_g + {}^2A_{1g} + {}^2A_{2g} + {}^2E_g + {}^2T_{1g} + {}^2T_{2g}\\ &= {}^4T_{1g} + {}^4T_{2g} + 2\,{}^2T_{1g} + 2\,{}^2T_{2g} + 2\,{}^2E_g + {}^2A_{1g} + {}^2A_{2g}\end{aligned}$$

This can be checked by consideration of the total degeneracy, since $(t_{2g})^4$ and $(e_g)^1$ have degeneracies of 15 and 4, respectively (Eqn. 12.1), so an overall degeneracy of 60 would be expected for $(t_{2g})^4(e_g)^1$, which is what is obtained $[12 + 12 + (2 \times 6) + (2 \times 6) + (2 \times 4) + 2 + 2]$.

Spin-allowed transitions from $(t_{2g})^4(e_g)^1 \leftarrow (t_{2g})^5$ would therefore be:

$$^2T_{1g} \leftarrow {}^2T_{2g} \qquad {}^2A_{1g} \leftarrow {}^2T_{2g} \qquad {}^2T_{2g} \leftarrow {}^2T_{2g}$$
$$^2A_{2g} \leftarrow {}^2T_{2g} \qquad {}^2E_g \leftarrow {}^2T_{2g}$$

In reality, the observed spectrum is far more complex than this, and is dominated by a number of intense charge transfer bands.

SAQ 13.8 :	*What spin-allowed transitions would be predicted for a low-spin d^6 complex $[(t_{2g})^5(e_g)^1 \leftarrow (t_{2g})^6]$?*

13.7 DESCENDING SYMMETRY

Both octahedral and tetrahedral complexes often have lower than ideal O_h / T_d symmetry. An example of where this occurs is an asymmetric distribution of ligands around the metal, as in the following examples:

$$\underset{(O_h)}{Mo(CO)_6} \rightarrow \underset{(D_{4h})}{\textit{trans}\text{-}Mo(CO)_4Cl_2} \rightarrow \underset{(C_{2v})}{\textit{cis}\text{-}Mo(CO)_4Cl_2}$$

$$\underset{(T_d)}{[CoCl_4]^{2-}} \rightarrow \underset{(C_{3v})}{[CoCl_3Br]^{2-}} \rightarrow \underset{(C_{2v})}{[CoCl_2Br_2]^{2-}}$$

A lowering of symmetry can also occur by a structural distortion, such as:

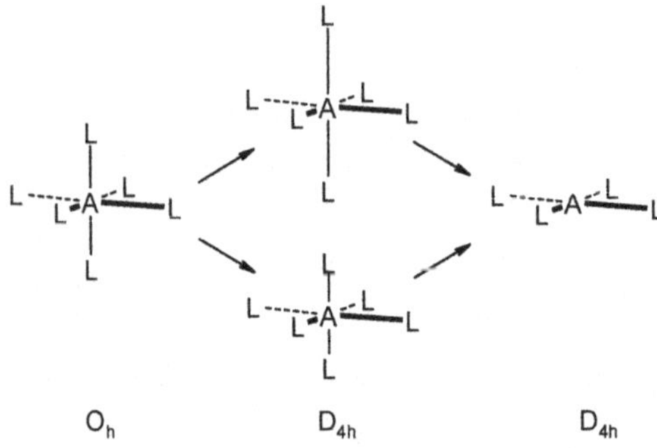

Either elongation or shortening of the axial bonds reduces O_h to D_{4h} symmetry; complete removal of the axial ligands generates a four-coordinate square-planar complex, also of D_{4h} symmetry. This **tetragonal distortion** is the basis of the **Jahn-Teller effect**, and it is this and its consequences for electronic spectra that will be the subject of this final section. The techniques described will, however, be applicable to other situations where symmetry is lowered across a series of related species e.g. changes to the υ(CO) vibrational modes of the three cobalt complexes mentioned above.

Comparison of the character tables for O_h and D_{4h} reveals a number of commonalities. Firstly, all the symmetry operations of the D_{4h} point group are contained in O_h and we can consider D_{4h} as one of its sub-groups. Some of the symmetry operations in O_h (C_3, S_6) are simply lost as symmetry is lowered, others change into more than one new operation:

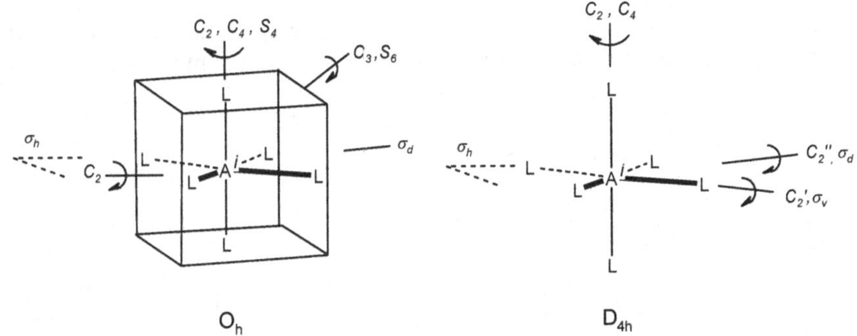

Each C_2 which is coincident with C_4 ($= C_4^2$) in O_h correlates with either C_2 or C_2' in D_{4h}, since two of the C_4 axes lie in horizontal plane (as drawn). The remaining C_2 axes in O_h (which bisect the ∠L-A-L) become C_2'', while the three σ_h planes become one σ_h and two σ_v in D_{4h}.

Table 13.3 Character table for O_h showing commonality with D_{4h} [a]

O_h	E	$6C_4$	$3C_2$[b]		$6C_2$	i	$6S_4$	$3\sigma_h$		$6\sigma_d$	
	E	$2C_4$	C_2	$2C_2'$	$2C_2''$	i	$2S_4$	σ_h	$2\sigma_v$	$2\sigma_d$	D_{4h}
A_{1g}	1	1	1	1	1	1	1	1	1	1	A_{1g}
A_{2g}	1	−1	1	1	−1	1	−1	1	1	−1	B_{1g}
E_g	2	0	2	2	0	2	0	2	2	0	$A_{1g} + B_{1g}$
T_{1g}	3	1	−1	−1	−1	3	1	−1	−1	−1	$A_{2g} + E_g$
T_{2g}	3	−1	−1	−1	1	3	−1	−1	−1	1	$B_{2g} + E_g$
A_{1u}	1	1	−1	1	1	−1	−1	−1	−1	−1	A_{1u}
A_{2u}	1	−1	−1	1	−1	−1	1	−1	−1	1	B_{1u}
E_u	2	0	2	2	0	−2	0	−2	−2	0	$A_{1u} + B_{1u}$
T_{1u}	3	1	−1	−1	−1	−3	−1	1	1	1	$A_{2u} + E_u$
T_{2u}	3	−1	−1	−1	1	−3	1	1	1	−1	$B_{2u} + E_u$

[a] The order in which operations are listed ($l - r$) are as in D_{4h}; the order for O_h has been rearranged to allow direct comparison between the two point group tables. [b] $= C_4^2$.

If we look at the characters of the irreducible representations for the common operations (Table 13.3) a correspondence between symmetry labels is evident. The table contains the symmetry operations common to both point groups and the irreducible representations for O_h and their characters are in the body of the table. Each row of characters represents one symmetry label in O_h (*left*) and a corresponding label in D_{4h} (*right*). For example, the characters in the irreducible representation for A_{2g} in O_h also represent B_{1g} in D_{4h}; we can say that A_{2g} in O_h **correlates** with B_{1g} in D_{4h}. A direct mapping is, however, not always possible. The characters representing irreducible E_g in O_h have no match in the D_{4h} character table,but are a reducible representation in this lower symmetry; that E_g (O_h) $\rightarrow$ A_{1g} + B_{1g} (D_{4h}) can be seen by adding the appropriate rows in Table 13.3, or by use of the reduction formula.

SAQ 13.9 : Confirm that T_{2u} (O_h) $\rightarrow B_{2u} + E_u$ (D_{4h})

Table 13.3 can be simplified to show the correlations as follows, in which the transformations of the symmetries of the *d*-orbitals have been emphasised:

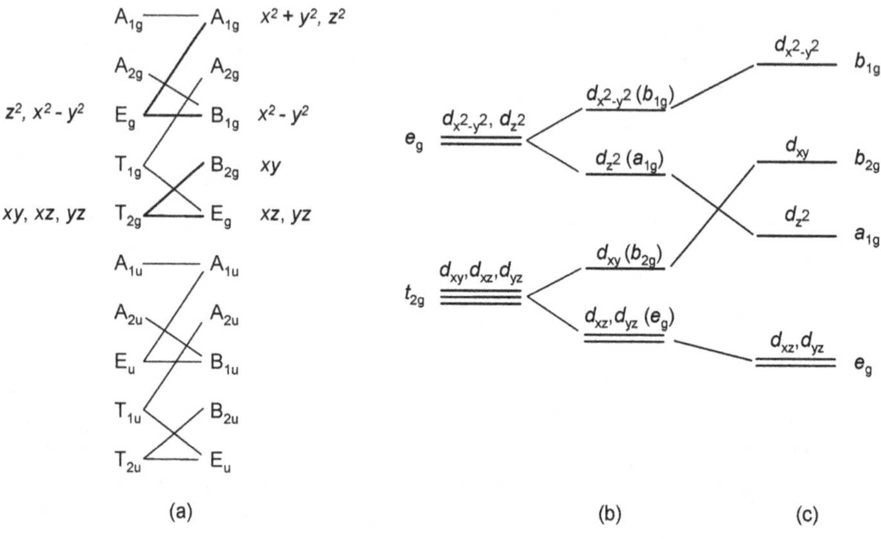

Fig. 13.6 (a) O_h / D_{4h} correlation diagram, and its implications for *d*-orbital spitting after (b) a tetragonal distortion and (c) complete removal of two ligands to give a square-planar complex.

In general, on descending symmetry properties which are degenerate have that degeneracy partially or completely removed. In the case of a tetragonal distortion which lowers O_h to D_{4h} symmetry, the t_{2g} orbital set splits into e_g and b_{2g}, while the e_g pair become a_{1g} and b_{1g}.

The Jahn-Teller theorem states that a non-linear degenerate system will distort to lower its symmetry and remove the degeneracy, though if an inversion centre is present in the degenerate state it is preserved on distortion. All d^n configurations save d^3, high-spin d^5, low spin d^6, d^8 and d^{10} (which have non-degenerate A symmetry), have associated E or T labels and will be susceptible to such a distortion. However, in practice only measureable distortions occur for doubly-degenerate E states i.e. when the e_g level has either one or three electrons. The driving force for removing degeneracy is the lowering of the energy of the system, which can be appreciated for d^9:

There is no net energy change from a redistribution of the $(t_{2g})^6$ electrons, but $(e_g)^3 \rightarrow (a_{1g})^2(b_{1g})^1$ results in a net lowering of energy for those in the e_g level. This occurs structurally by elongation of two opposite bonds in the octahedral complex (usually taken to be those along *z*) while preserving the inversion centre,† lowering the symmetry to D_{4h}. In a simple crystal field analysis, the d_{z^2} orbital now experiences less repulsion from the ligands along *z* so is lowered in energy, while the $d_{x^2-y^2}$ orbital increases its energy (it is now the only one which points directly at the nearby ligands) by an equal extent, to preserve the overall energy of the five *d*-orbitals. As two electrons benefit from the lowering of the d_{z^2} orbital energy and only one increases its energy as $d_{x^2-y^2}$ is destabilised there is a net energy lowering. Similar arguments apply to high spin d^4 $[(t_{2g})^3(e_g)^1]$ and low spin d^7 $[(t_{2g})^6(e_g)^1]$.

Triply-degenerate configurations e.g. d^1 $[(t_{2g})^1]$ will undergo smaller distortions as the distinction between the three t_{2g} orbitals, none of which point directly at any of the six ligands, is less marked then for the e_g orbital pair. Such distortions are usually hard to detect experimentally.

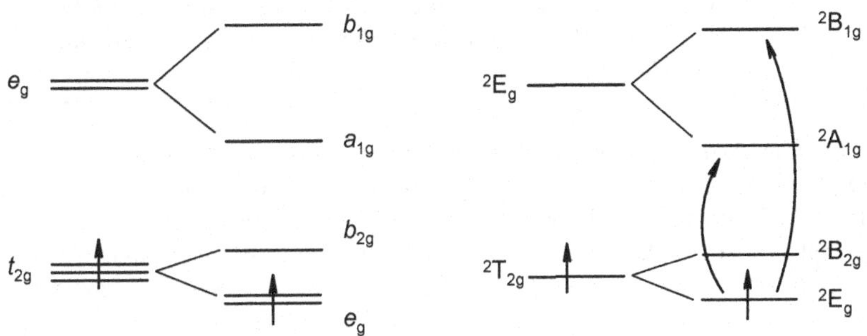

The consequences of the Jahn-Teller effect in electronic spectroscopy are a splitting of bands as orbital degeneracy in removed. The d^1 system is a case in point (*above*), exemplified by the purple aqua-anion $[Ti(H_2O)_6]^{3+}$ which exhibits a single broad band at 20,300 cm^{-1} clearly split into two by *ca.* 2,000 cm^{-1}. The d^1 case, is however somewhat more complex than it might appear because a Jahn-Teller splitting of the $(t_{2g})^1$ ground state is not expected to be significant enough to generate such a large band separation. In fact, the splitting of the band arises due to a tetragonal distortion of the excited $(e_g)^1$ state. The terms associated with the $(t_{2g})^1$ and $(e_g)^1$ configurations

† A contraction of the two bonds lying along *z* and an elongation of the four in the *xy* plane also results in D_{4h} symmetry, though this is far less commonly observed.

are $^2T_{2g}$ and ^2E_g, respectively, and these split on lowering the symmetry from O_h to D_{4h} in the same way that *d*-orbital degeneracy is lowered: The two absorptions which give rise to the spectrum are $^2A_{1g} \leftarrow {}^2E_g$ and $^2B_{1g} \leftarrow {}^2E_g$, and the band splitting corresponds to the $^2A_{1g}$ / $^2B_{1g}$ energy difference.

Jahn-Teller distortions of d^9-sytems such as Cu(II) are among the most commonly observed and can be most easily rationalised in terms of the hole formalism. The ^2D ground term for d^9 under O_h symmetry will be split as follows on symmetry descending to D_{4h}:

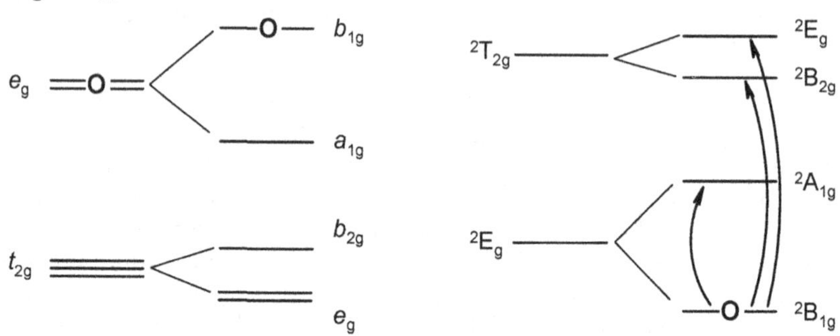

The d^9 configuration can be viewed as a hole (o) in the e_g level, which on distortion to D_{4h} will change to b_{1g} i.e. specifically the $d_{x^2-y^2}$ orbital. Similary, the ground term becomes $^2B_{1g}$ (splitting has no impact on spin multiplicity), and the observed electronic spectrum – a broad envelope with a maximum at *ca.* 12,000 cm^{-1} – rationalised by overlapping $^2A_{1g} \leftarrow {}^2B_{1g}$ (the hole moving to d_{z^2}) and unresolved $^2B_{2g} \leftarrow {}^2B_{1g}$ / $^2E_g \leftarrow {}^2B_{1g}$ transitions.

SAQ 13.10 : By analogy with the d^9 case, suggest electronic transitions which explain the broad band at ca. 14,000 cm^{-1} in the electronic spectrum of $[Cr(H_2O)_6]^{2+}$.

This analysis of descending symmetry is not confined to energy levels and electronic spectra alone but applies to any of the properties of a system. The vibrational spectra which have been analysed in Part II of this book are equally amenable to such a treatment. For example, how does the vibrational spectrum of CH_4 (T_d) change in terms of the number of observed bands when the symmetry is lowered to C_{3v} (e.g. CH_3D) or C_{2v} (e.g. CH_2D_2). For CH_4, using the standard techniques, group theory predicts:

$$\Gamma_{vib} = A_1 + E + 2T_2$$

The following correlation table links $T_d \rightarrow C_{3v} \rightarrow C_{2v}$:

$T_d \rightarrow$	C_{3v}	C_{2v}
A_1	A_1	A_1
A_2	A_2	A_2
E	E	$A_1 + A_2$
T_1	$A_2 + E$	$A_2 + B_1 + B_2$
T_2	$A_1 + E$	$A_1 + B_1 + B_2$

Using the above table, the vibrational spectra for the lower symmetry deuterated species would be:

CH_3D ($T_d \rightarrow C_{3v}$), $\Gamma_{vib} = A_1 + E + 2(A_1 + E) = 3A_1 + 3E$ (*see Chapter 5, problem 2*)

CH_2D_2 ($T_d \rightarrow C_{2v}$), $\Gamma_{vib} = A_1 + (A_1 + A_2) + 2(A_1 + B_1 + B_2)$

$$= 4A_1 + A_2 + 2B_1 + 2B_2$$

These predictions are borne out by experiment.

13.8 SUMMARY

- the Beer-Lambert law states $A = \varepsilon \times l \times C$.
- the Laporte selection rule requires a symmetry-allowed transition to have $\Delta l = \pm 1$.
- the parity rule forbids $g \rightarrow g$ and $u \rightarrow u$ transitions
- both the parity and Laporte rules are broken by vibronic coupling, when the symmetry of a molecular vibration matches (spans), in part, that of the electronic transition integral.
- the spin selection rule is violated by spin-orbit coupling, but the intensity of bands arising from such transitions is weak.
- terms for high-spin complexes have $d^n \equiv d^{n\pm5}$ ($d^1 \equiv d^6$, $d^2 \equiv d^7$, $d^3 \equiv d^8$, $d^4 \equiv d^9$), for both octahedral and tetrahedral cases.
- terms for octahedral $d^n \equiv$ tetrahedral d^{10-n} and *vice versa*.

Octahedral		Tetrahedral	Ground term
d^1, d^6	$\equiv$	d^4, d^9	*D*
d^4, d^9	$\equiv$	d^1, d^6	*D*
d^2, d^7	$\equiv$	d^3, d^8	*F*
d^3, d^8	$\equiv$	d^2, d^7	*F*

- in a ligand field, splitting of terms for d^n are the reverse of d^{10-n} (e.g. d^1, d^9).
- low-spin complexes have a ground term which is not of that of maximum multiplicity.
- the Jahn-Teller theorem states that a degenerate system will distort to lower its symmetry and remove the degeneracy.

PROBLEMS

Answers to all problems marked with * are given in Appendix 4.

1*. For a d^1 species, by taking the appropriate direct products evaluate the symmetry of the transition integral to show that the $^2T_2 \leftarrow {}^2E$ transition is allowed for a tetrahedral species but $^2E_g \leftarrow {}^2T_{2g}$ is symmetry-forbidden for an octahedral complex.

2. Show that the lowest energy electronic transition for an octahedral d^2 complex, $^3T_{2g} \leftarrow {}^3T_{1g}(F)$, is symmetry forbidden but vibronically allowed.

3*. Low-spin Co^{3+} species show two low energy bands in their electronic spectra corresponding to a transtion from $(t_{2g})^6$ to $(t_{2g})^5(e_g)^1$. Rationalise these two bands in terms of the symmetries of the ground and excited states.

4*. The ground term for a low-spin d^4 ion $[(t_{2g})^4]$ in an octahedral field is $^3T_{1g}$. Using Table 12.6 in Section 12.5, determine the symmetry labels for all the terms associated with the first excited state $[(t_{2g})^3(e_g)^1]$ and check the degeneracy of your answer is correct. Then, predict the spin-allowed transitions for low-spin d^4 to its first excited configuration.

5. The d^8 ion $[Pt(CN)_4]^{2-}$ adopts a square planar structure (D_{4h}) which has a HOMO of b_{2g} symmetry and a LUMO which is b_{1g} (Fig. 13.6c).

   (a) What are the symmetries of the ground and first excited states ?
   (b) Show that a transition between these states is symmetry-forbidden.
   (c) Given that $\Gamma_{vib} = A_{1g} + B_{1g} + B_{2g} + A_{2u} + B_{2u} + 2E_u$, is the transition vibronically allowed?

APPENDICES

APPENDIX 1
Projection Operators

In Chapter 6 (*Section 6.3*) the limitations of a pictorial approach to generating MOs was touched on. A SALC is a linear combination of atomic orbitals:

$$\text{SALC } \psi = N[c_1\psi_1 + c_2\psi_2 + c_3\psi_3 \ldots\ldots] \qquad \text{(eqn. A.1)}$$

where N is a normalisation constant and ψ_i are the wavefunctions for individual AOs. In the qualitative approach to describing MOs adopted throughout this book the coefficients (c_i) associated with each AO contribution are ignored (i.e. assumed to be equal). Although this does not affect significantly the pictorial description of the MO, it lacks a level of detail which can, at times, lead to problems. It has already been commented on that these coefficients are not equal when the AO energies are different, but, in addition, symmetry also dictates that they need not necessarily be equal even when the AO energies are the same.

This Appendix will set out the rigorous approach that is required to generate exact forms of the SALCs using a technique called **projection operators**. This is a difficult topic and only the basic methodology will be described. In addition to its application to MO theory, the same technique can be used to generate the forms of the vibrational modes modes discussed in Part 2 of this book (*see Section 5.1*), which are themselves SALCs, but now linear combinations of individual vibrational modes.

A.1 THE BASICS – NON-DEGENERATE SALCs

In Section 7.1 we saw that the SALCs for the two hydrogen atoms in water (C_{2v}) could be described by the irreducible representations $a_1 + b_2$. Since there are only two ways in which the two hydrogen AOs can combine it was trivial to draw these possibilities:

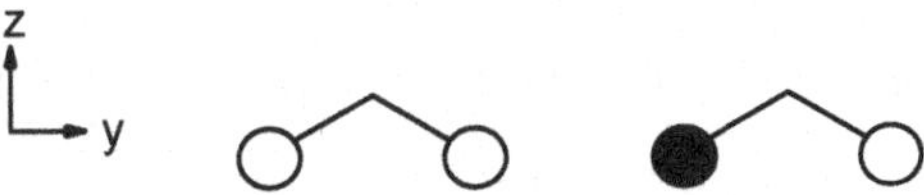

How can we arrive at these in a more formal way, so that more difficult situations can be analysed? We do this using a projection operator technique, which is essentially a mechanism for automatically generating algebraic functions such as in equation A.1. As with the reduction formula, the appearance of the projection operator formula looks intimidating but is relatively easy to apply:

$$P_i\psi_j = \sum_R (\chi_{(IR)} R\psi_j) \quad \text{sum over all classes of operation R} \qquad \text{(eqn. A.2)}$$

P_i is the projection operator for a given irreducible representation
ψ_j is the chosen basis (or generating) function
$\chi_{(IR)}$ is the character in the irreducible representation for operation R
$R\psi_j$ is the effect of performing operation R on the basis function ψ_j

This rather daunting equation says "the result of applying the projection operator P_i to a chosen basis function ψ_j is arrived at by summing, over all operations R in the point group, the product of the character in the irreducible representation for the operation $\chi_{(IR)}$ and the effect of the operation on the basis function"! An example will illustrate this is not as difficult as it sounds.

Let us take one of the H 1s orbitals as our basis function, ψ_1:

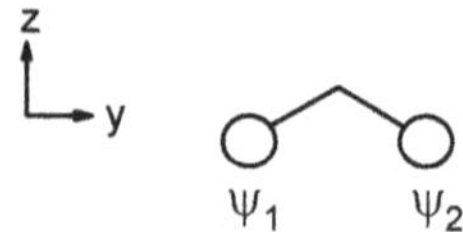

We need to find the SALCs associated with the a_1 and b_2 representations, so for a_1:

C_{2v}	E	C_2	$\sigma(xz)$	$\sigma(yz)$	
$\psi_1 \rightarrow$	ψ_1	ψ_2	ψ_2	ψ_1	
$\chi_{(IR)}\, a_1$	1	1	1	1	
$\chi_{(IR)} R\psi_1$	ψ_1	ψ_2	ψ_2	ψ_1	$= 2\psi_1 + 2\psi_2$

Since we are only interested in the relative contributions of the two AOs to the MO, this can be simplified to:

$$a_1 = \psi_1 + \psi_2$$

which is what is expected for the in-phase combination of AOs that make up the bonding MO.

The SALC needs finally to be normalised to be completely accurate, so that the total electron density sums to one electron. Since the probability of finding an electron at any point is given by ψ^2, we need a normalisation constant N which reduces the sum of the squares of the coefficients (c_i in equation A.1) to unity. Since, for the a_1 SALC, $c_1 = c_2 = 1$, $N = 1/\sqrt{2}$.

i.e. $a_1 = (1/\sqrt{2})\psi_1 + (1/\sqrt{2})\psi_2 = 1/\sqrt{2}(\psi_1 + \psi_2)$, so that $(1/\sqrt{2})^2 + (1/\sqrt{2})^2 = 1$

In general, $N = 1/\sqrt{n}$, where $n = \Sigma(c_i)^2$.

SAQ A.1 : Using the projection operator method with ψ_1 as basis function, show that the MO of b_2 symmetry in H_2O is of the form $1/\sqrt{2}(\psi_1 - \psi_2)$.

Answers to all SAQs are given in Appendix 3.

A final check to make sure the SALCs that have been generated are correct is that they must be **orthogonal** to each other. The mathematics of this need not concern us, we simply note how it is done:

- functions are orthogonal if the sum of the products of corresponding coefficients is zero i.e. product of coefficients of ψ_1 + product of coefficients of ψ_2 ..etc.

For the two MO SALCs of water this is simply $(1 \times 1) + (1 \times -1) = 0$

A.2 DEGENERATE SALCs

Problem 1 at the end of Chapter 8 looked at the MO diagram for the π-bonds in BF_3 (D_{3h}).

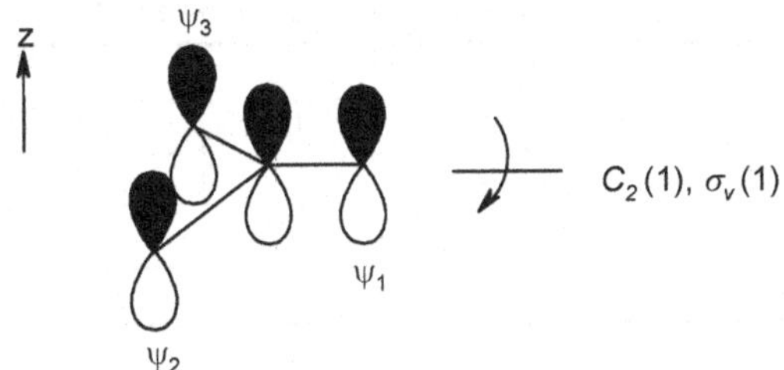

The p_z orbital on boron has a_2'' symmetry, while the SALCs associated with the three p_z orbitals on fluorine transform as a_2'' + e'' (*see Appendix 4, Chapter 8, Problem 1*). The form of these SALCs can easily be drawn using the methodology of Chapter 8 and *SAQ* 8.2:

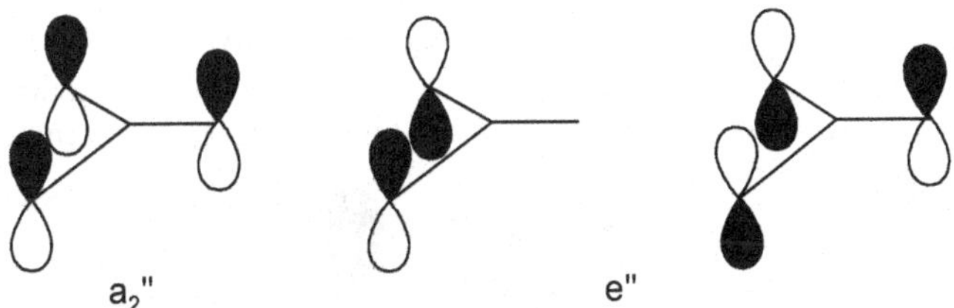

The match beween the fluorine SALC and boron AO each of a_2'' symmetry is clear, but close consideration of the mismatch between this AO on boron and the e'' fluorine SALCs raises a problem: the p_z on boron has equal amounts of bonding and anti-bonding overlap with one SALC making it a non-bonding interaction overall, but with the other SALC there appears more bonding than anti-bonding, which cannot be correct. A consideration of the exact form of the SALC will rationalise this apparent anomaly.

We begin by applying the projection operator method using ψ_1 as the generating function. In seeing how this transforms under the operations of the point group, we must consider all the operations of the point group individually.

For the a_2'' SALC:

D_{3h}	E	C_3^1	C_3^2	$C_2(1)$	$C_2(2)$	$C_2(3)$	σ_h
$\psi_1 \rightarrow$	ψ_1	ψ_2	ψ_3	$-\psi_1$	$-\psi_3$	$-\psi_2$	$-\psi_1$
$\chi_{(IR)}\, a_2''$	1	1	1	−1	−1	−1	−1
$\chi_{(IR)} R\psi_1$	ψ_1	ψ_2	ψ_3	ψ_1	ψ_3	ψ_2	ψ_1

...continued

D_{3h}	S_3^1	S_3^5	$\sigma_v(1)$	$\sigma_v(2)$	$\sigma_v(3)$
$\psi_1 \rightarrow$	$-\psi_2$	$-\psi_3$	ψ_1	ψ_3	ψ_2
$\chi_{(IR)}\, a_2''$	−1	−1	1	1	1
$\chi_{(IR)} R\psi_1$	ψ_2	ψ_3	ψ_1	ψ_3	ψ_2

Note that the "$2S_3$" operations in the shortened form of the character table refer to S_3^{1} and S_3^{5}; $S_3^{2} \equiv C_3^{2}$, $S_3^{3} \equiv \sigma_h$, $S_3^{4} \equiv C_3^{1}$ and $S_3^{6} \equiv E$.

Thus, $a_2'' = 4\psi_1 + 4\psi_2 + 4\psi_3$, or $a_2'' = \psi_1 + \psi_2 + \psi_3$ considering only the relative contributions of each AO. Including the normalisation coefficient, this becomes $1\sqrt{3}(\psi_1 + \psi_2 + \psi_3)$. By the same method, for the e'' SALC:

D_{3h}	E	C_3^{1}	C_3^{2}	$C_2(1)$	$C_2(2)$	$C_2(3)$	σ_h
$\psi_1 \rightarrow$	ψ_1	ψ_2	ψ_3	$-\psi_1$	$-\psi_3$	$-\psi_2$	$-\psi_1$
$\chi_{(IR)}\, e''$	2	-1	-1	0	0	0	-2
$\chi_{(IR)}R\psi_1$	$2\psi_1$	$-\psi_2$	$-\psi_3$				$2\psi_1$

...continued

D_{3h}	S_3^{1}	S_3^{5}	$\sigma_v(1)$	$\sigma_v(2)$	$\sigma_v(3)$
$\psi_1 \rightarrow$	$-\psi_2$	$-\psi_3$	ψ_1	ψ_3	ψ_2
$\chi_{(IR)}\, e''$	1	1	0	0	0
$\chi_{(IR)}R\psi_1$	$-\psi_2$	$-\psi_3$			

$$e'' = 4\psi_1 - 2\psi_2 - 2\psi_3 \text{ or } e'' = 2\psi_1 - \psi_2 - \psi_3 \text{ [normalised : } 1/\sqrt{6}(2\psi_1 - \psi_2 - \psi_3)]$$

Pictorially, we should redraw the second e'' SALC showing the double contribution from ψ_1, and which now has equal amounts of bonding and anti-bonding overlap with the p_z on boron making the interaction, correctly, non-bonding:

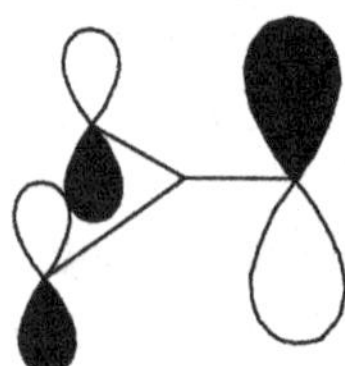

SAQ A.2 : Show that the two SALCs derived above of a_2'' and e'' symmetry are orthogonal.

While this is one problem resolved, another has been raised. Using ψ_1 as generating function has only yielded the form of one of the two SALCs of e'' symmetry. To generate the form of the other half of this degenerate pair we need a new generating function and this is where the use of projection operators becomes less intuitive.

It might seem reasonable to try either ψ_2 or ψ_3 as an alternative generating function, but neither of these gives acceptable answers. Using ψ_2 as generating function, and ignoring those operations for which $\chi_{(IR)}$ is zero for e'':

D_{3h}	E	C_3^{1}	C_3^{2}	σ_h	S_3^{1}	S_3^{5}
$\psi_2 \rightarrow$	ψ_2	ψ_3	ψ_1	$-\psi_2$	$-\psi_3$	$-\psi_1$
$\chi_{(IR)}\, e''$	2	-1	-1	-2	1	1
$\chi_{(IR)}R\psi_1$	$2\psi_2$	$-\psi_3$	$-\psi_1$	$2\psi_2$	$-\psi_3$	$-\psi_1$

This gives $e'' = 4\psi_2 - 2\psi_3 - 2\psi_1$ or $e'' = 2\psi_2 - \psi_3 - \psi_1$ (ignoring normalisation). However, this is not orthogonal to the other e'' function, as multiplying common coefficients for ψ_1, ψ_2 and ψ_3 shows :

$$(2 \times -1) + (-1 \times 2) + (-1 \times -1) = -3, \text{ not 0 as required for othogonality.}$$

SAQ A.3 :	*Determine the e'' function that is generated using ψ_3 as generator and show that this is not orthogonal to the function generated for e'' using ψ_1.*

The remaining e'' function can be arrived at using $\psi_2 - \psi_3$ as the generating function. It is not at all intuitive that this is the generator to chose, and this is one of the limitations of the projection operator method at a basic level of group theory. Rational ways of determining the choice of basis functions for degenerate MOs (and vibrational modes) i.e.those of *e* or *t* symmetry, can be found in more advanced texts on the subject.†

Using $\psi_2 - \psi_3$ as generator and the same methodology as before gives:

D_{3h}	E	C_3^1	C_3^2	σ_h	S_3^1	S_3^5
$\psi_2 - \psi_3 \rightarrow$	$\psi_2 - \psi_3$	$\psi_3 - \psi_1$	$\psi_1 - \psi_2$	$-\psi_2 + \psi_3$	$-\psi_3 + \psi_1$	$-\psi_1 + \psi_2$
$\chi_{(IR)}\, e''$	2	-1	-1	-2	1	1
$\chi_{(IR)} R\psi_1$	$2\psi_2 - 2\psi_3$	$-\psi_3 + \psi_1$	$-\psi_1 + \psi_2$	$2\psi_2 - 2\psi_3$	$-\psi_3 + \psi_1$	$-\psi_1 + \psi_2$

From which:

$$e'' = 6\psi_2 - 6\psi_3 \text{ or } e'' = \psi_2 - \psi_3 \ [1/\sqrt{2}(\psi_2 - \psi_3) \text{ including normalisation}]$$

This is both orthogonal to the other e'' SALC $[(2 \times 0) + (-1 \times 1) + (-1 \times -1) = 0]$ and consistent with the pictorial representation of the SALC shown at the beginning of this Section.

A.3 VIBRATIONAL MODES

We can also use projection operators to generate the functions which represent modes in a vibrational spectrum, as these can be considered as linear combinations of individual stretching or bending modes, in a manner dictated by symmetry, just as SALCs of AOs have been derived above.

Using the methodology described in Chapters 3 and 5, the vibrational modes of NH_3 (C_{3v}) can be shown to be:

$$\Gamma_{vib} = 2A_1 + 2E$$
$$\Gamma_{N\text{-}H} = A_1 + E$$
$$\Gamma_{bend} = A_1 + E$$

---

† For example:
Molecular Symmetry and Group Theory, R L Carter, John Wiley and Sons, 1998.
Chemical Applications of Group Theory 3rd Edition, F A Cotton, John Wiley and Sons, 1990.

We can use the vectors v_1, v_2, v_3 to generate the forms of the stretching modes and the angles α_1, α_2, α_3 for the bends:

For the stretching modes, using v_1 as generator gives:

C_{3v}	E	C_3^1	C_3^2	$\sigma_v(1)$	$\sigma_v(2)$	$\sigma_v(3)$	
$v_1 \rightarrow$	v_1	v_2	v_3	v_1	v_3	v_2	
$\chi_{(IR)}$ A_1	1	1	1	1	1	1	
$\chi_{(IR)}Rv_1$	v_1	v_2	v_3	v_1	v_3	v_2	$\rightarrow 1/\sqrt{3}(v_1 + v_2 + v_3)$
$\chi_{(IR)}$ E	2	−1	−1	0	0	0	
$\chi_{(IR)}Rv_1$	$2v_1$	$-v_2$	$-v_3$				$\rightarrow 1/\sqrt{6}(2v_1 - v_2 - v_3)$

The second component of the E mode requires $v_2 - v_3$ as generating function:

C_{3v}	E	C_3^1	C_3^2	
$v_2 - v_3 \rightarrow$	$v_2 - v_3$	$v_3 - v_1$	$v_1 - v_2$	
$\chi_{(IR)}$ E	2	−1	−1	
$\chi_{(IR)}Rv_1$	$2v_2 - 2v_3$	$-v_3 + v_1$	$-v_1 + v_2$	$\rightarrow 1/\sqrt{2}(v_2 - v_3)$

Pictorially, these appear as:

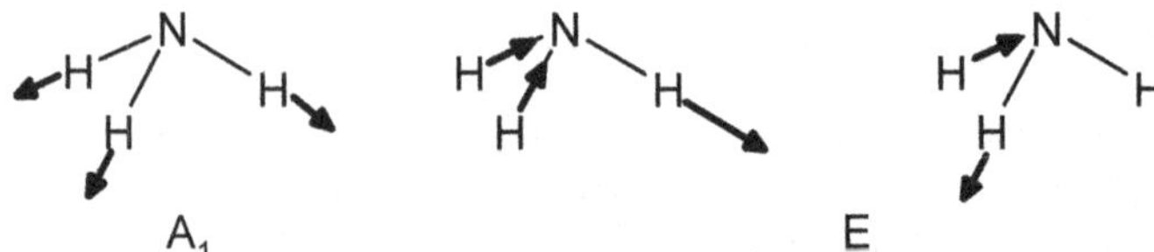

For completeness, in each case the central nitrogen will also move to keep the molecule stationary, but this has been omitted for clarity of the stretching motions, both above and in further examples in this Appendix.

SAQ A.4 : *Use α_1 and $\alpha_2 - \alpha_3$ as generators to determine the functions which describe the bending modes for NH_3.*

There are clear analogies between the forms of these vibrational SALCs and the SALCs for cyclic combinations of AOs (*Chapter 8*). The A_1 bending mode ($\alpha_1 + \alpha_2 + \alpha_3$) requires all angles to behave the same (the "+" in Figure A.1 shows them all expanding at the same time) which occurs by each hydrogen moving up out of the H_3 plane (the molecule becomes less pyramidal). The first of the E bends has one angle expanding and two contracting, analogous to the AO SALC in which two AOs are in-phase but out-of-phase with the third. The second E bend has one angle expanding at the expense of one other, the third angle remaining static.

AO SALCs

stretching modes

bending modes

Figure A.1. Hydrogen SALCs and vibrational modes for NH_3.

Thus, without recourse to projection operators the method for generating cyclic SALCs from Chapter 8 also offers a simplistic way of depicting the vibrational modes.

A final example, referring back to our analysis of the vibrational spectrum of $[PtCl_4]^{2-}$ (*Chapter 5*), will reinforce this, but also highlight that this pictorial approach too has its limitations. From a knowledge of how the four SALCs for a cyclic array of AOs are arranged (*SAQ* 8.2), we would predict the following for the vibrational modes of $[PtCl_4]^{2-}$:

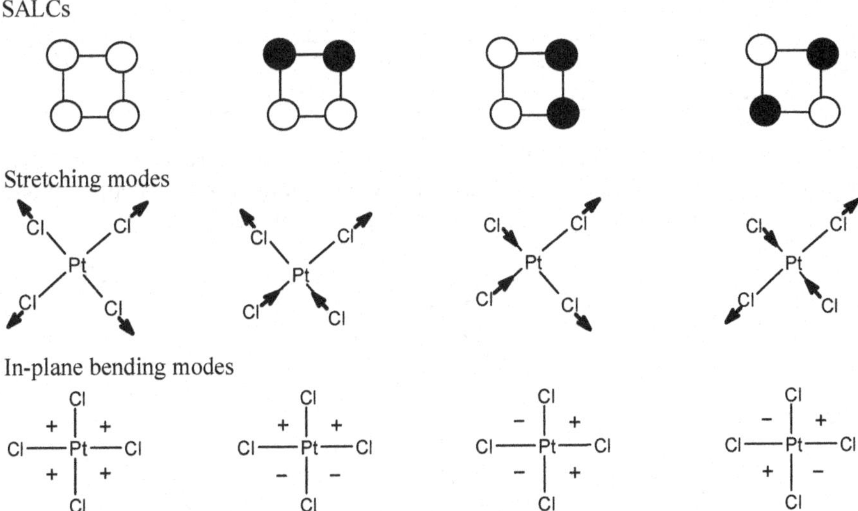

Out-of-plane bending modes

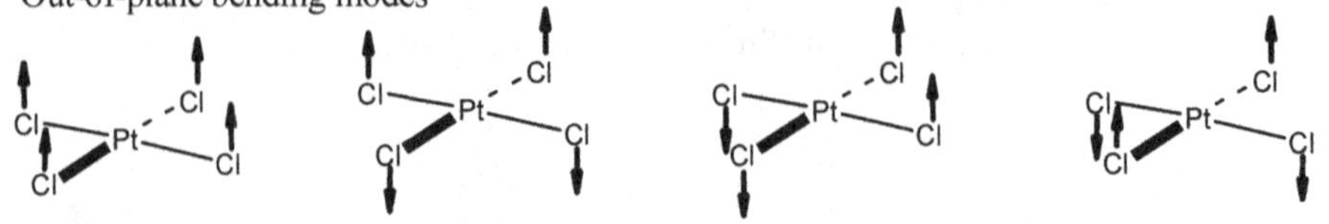

Figure A.2. Predicted vibrational modes for $[PtCl_4]^{2-}$.

From Chapter 5 (including *SAQ* 5.1), the vibrational modes for $[PtCl_4]^{2-}$ were shown to be (redundancies in parentheses):

$$\Gamma_{vib} = A_{1g} + B_{1g} + B_{2g} + A_{2u} + B_{2u} + 2E_u$$

$$\Gamma_{Pt\text{-}Cl} = A_{1g} + B_{1g} + E_u$$

$$\Gamma_{in\text{-}plane} = (A_{1g}) + B_{2g} + E_u$$

$$\Gamma_{out\text{-}of\text{-}plane} = (E_g) + A_{2u} + B_{2u}$$

The functions corresponding to the stretching modes can be determined simply from the projection operator method.

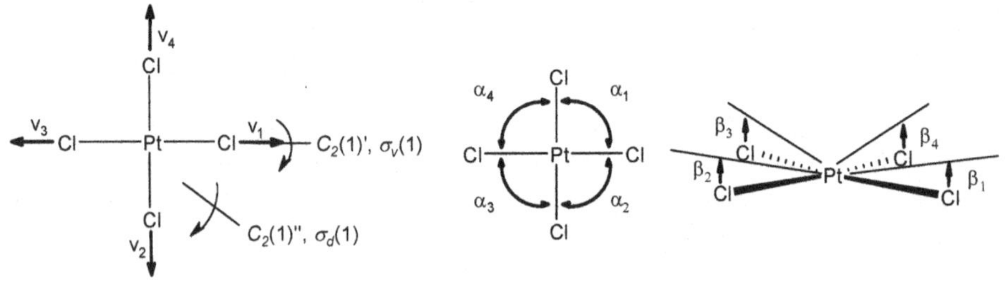

SAQ A.5 : *Use v_1 (for A_{1g}, B_{1g}, E_u) and v_2 (for E_u) as generators to determine the functions which describe the stretching modes for $[PtCl_4]^{2-}$.*

The A_{1g} ($v_1 + v_2 + v_3 + v_4$) and B_{1g} ($v_1 - v_2 + v_3 - v_4$) modes are as in Figure A.2, while the E_u modes ($v_1 - v_3$; $v_2 - v_4$) are simply the symmetry-equivalent sum and difference pairs from those drawn (*see Section 9.1* for a related case):

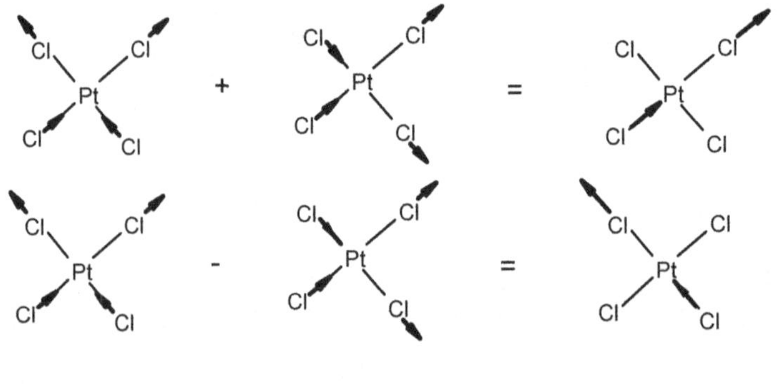

The out-of plane bending modes are arrived at in an analogous manner, using angles β_1 and β_2 as generating functions.

SAQ A.6 : Use β_1 (for A_{2u}, B_{2u}, E_g) and β_2 (for E_g) as generators to show the functions which describe the out-of-plane bending modes for $[PtCl_4]^{2-}$ are:

$$A_{2u} = 1/2(\beta_1 + \beta_2 + \beta_3 + \beta_4)$$

$$B_{2u} = 1/2(\beta_1 - \beta_2 + \beta_3 - \beta_4)$$

$$E_g = 1/\sqrt{2}(\beta_1 - \beta_3),\ 1/\sqrt{2}(\beta_2 - \beta_4)$$

Note that the E_g redundancy among the out-of-plane modes is because these are, in fact, rotations about *x* and *y* (Fig. A.2).

Deriving functions for the in-plane bending modes has the recurrent difficulty for this basic description of projection operator methods of choosing the correct generating function. While α_1 and α_2 would seem logical based on the methodology of *SAQ* A.5 and *SAQ* A.6, these are not completely acceptable. They can be used as generators for the singly-degenerate A_{1g} and B_{2g} modes, but not for the degenerate E_u in-plane bends. This is because it is a requirement of the generating functions for degenerate systems that they all have the same symmetry, and in this respect α_1 and α_2 differ from v_1 and v_2 (or β_1, β_2) used to derive the functions for the degenerate E_u stretching and E out-of-plane bending modes. For example, under $C_2'(1)$, $v_1 \rightarrow v_1$ (i.e. itself) but $\alpha_1 \rightarrow \alpha_4$ (not itself). Functions $(\alpha_1 + \alpha_2)$ and $(\alpha_1 + \alpha_4)$ do meet these requirements [e.g. $C_2'(1)$, $(\alpha_1 + \alpha_4) \rightarrow (\alpha_1 + \alpha_4)$] and the interested reader may wish to show that using this combination of functions generates the following, consistent with Figure A.2:

$$A_{1g} = 1/2(\alpha_1 + \alpha_2 + \alpha_3 + \alpha_4)$$

$$B_{2g} = 1/2(\alpha_1 - \alpha_2 + \alpha_3 - \alpha_4)$$

$$E_u = 1/\sqrt{2}(\alpha - \alpha_3),\ 1/\sqrt{2}(\alpha_2 - \alpha_4)$$

This appendix concludes with a note of caution. Depicting molecular vibrations using combined qualitative approach in parallel with the use of projection operators is, in itself, not without limitations. We have already seen in the platinum example that apparently different, but symmetry-equivalent, SALCs can be derived from the two methods. This is usually a problem only with doubly- and triply-degenerate species (E, T labels), and it is up to the user to make use of the most appropriate form for the problem being addressed.

APPENDIX 2
Microstates and Term Symbols

This brief overview sets out a stepwise strategy for determining the microstates for a given electronic configuration and how these are divided up into groups, each described by a term symbol. The microstates and term symbols for the d^2 configuration will be used as an example.

Step 1* : *Determine the degeneracy of the configuration

This is done using eqn.12. 1:

$$D_t = (N)! / (N_e)!(N_h)!$$

$$D_t = (2 \times 5)! / 2! \times 8! \;=\; 10 \times 9 / 2 \;= 45$$

Step 2* : *Determine the maximum L and S for the configuration

For each *d*-electron, $m_l = 2, 1, 0, -1, -2$, so maximum $L = 4$ (2 + 2). The two electrons can both have $m_s = ½$ so maximum $S = 1$.

Step 3* : *Determine the range of M_L and M_S values associated with maximum L, S

$L = 4$ has associated M_L values of 4, 3, 2, 1, 0, −1, −2, −3, −4; $S = 1$ has associated M_S values of 1, 0, −1. The range of possible microstates must lie within a grid which spans these M_L / M_S ranges (Table A.1).

Step 4* : *Generate and complete a grid of microstates

The maximum M_L (= 4) arises from $\Sigma m_l = 2 + 2$. As the two electrons are in the same *d*-orbital their spins must be paired given the table entry $(2^+,2^-)$. The next highest possible $M_L = 3$, and arises from $\Sigma m_l = 2 + 1$. As the two electrons are in different orbitals they can take either spin, giving rise to four microstates $(2^+,1^+)$, $(2^+,1^-)$, $(2^-,1^+)$, $(2^-,1^-)$.

This process can be followed upto and including the row for $M_L = 0$, after which point the entries for $M_L = -1, -2, -3, -4$ mirror those for $M_L = 1, 2, 3, 4$. The complete grid of microstates is given in Table A.1.

Step 5* : *Identify the terms and their associated microstates from the grid

At this point it is worth copying Table A.1 so that the microstates can be crossed off as each term is identified. The first entry in the Table is $(2^+,2^-)$ corresponding to $M = 4$ and $M_S = 0$. This is one microstate that is part of the ^1G term ($L = 4 \Rightarrow G$; $S = 0 \Rightarrow$ singlet) with an overall degeneracy of $(2L+1)(2S+1) = 9$. The remaining eight microstates all have $S = 0$ (this is the only value associated with $M_S = 0$) and $M_L = 3$, 2, 1, 0, −1, −2, −3, −4 and are shown shaded in the partial grid of Table A.2 e.g. $(2^+,1^-)$, though it doesn't matter which entry in any given M_L / M_S square is removed. These microstates should now be crossed off from Table A.1

Of the microstates remaining the one with highest L and S is $(2^+,1^+)$, with $L = 3$ and $S = 1$; this is part of a ^3F term ($L = 3 \Rightarrow F$; $S = 1 \Rightarrow$ triplet). ^3F has a degeneracy

Table A.1 Microstate table for a d^2 configuration.

M_L	M_S = 1	M_S = 0	M_S = -1
4		$(2^+,2^-)$	
3	$(2^+,1^+)$	$(2^+,1^-)$ $(2^-,1^+)$	$(2^-,1^-)$
2	$(2^+,0^+)$	$(2^+,0^-)$ $(2^-,0^+)$ $(1^+,1^-)$	$(2^-,0^-)$
1	$(2^+,-1^+)$ $(1^+,0^+)$	$(2^+,-1^-)$ $(2^-,-1^+)$ $(1^+,0^-)$ $(1^-,0^+)$	$(2^-,-1^-)$ $(1^-,0^-)$
0	$(2^+,-2^+)$ $(1^+,-1^+)$	$(2^+,-2^-)$ $(2^-,-2^+)$ $(1^+,-1^-)$ $(1^-,-1^+)$ $(0^+,0^-)$	$(2^-,-2^-)$ $(1^-,-1^-)$
-1	$(-2^+,1^+)$ $(1^+,0^+)$	$(-2^+,1^-)$ $(-2^-,1^+)$ $(-1^+,0^-)$ $(-1^-,0^+)$	$(-2^-,1^-)$ $(-1^-,0^-)$
-2	$(-2^+,0^+)$	$(-2^+,0^-)$ $(-2^-,0^+)$ $(-1^+,-1^-)$	$(-2^-,0^-)$
-3	$(-2^+,-1^+)$	$(-2^+,-1^-)$ $(-2^-,-1^+)$	$(-2^-,-1^-)$
-4		$(-2^+,-2^-)$	

Table A.2 Partial microstate table for a d^2 configuration highlighting some of the ^1G microstates.

M_L	M_S = 1	M_S = 0	M_S = -1
4		$(2^+,2^-)$	
3	$(2^+,1^+)$	$(2^+,1^-)$ $(2^-,1^+)$	$(2^-,1^-)$
2	$(2^+,0^+)$	$(2^+,0^-)$ $(2^-,0^+)$ $(1^+,1^-)$	$(2^-,0^-)$
1	$(2^+,-1^+)$ $(1^+,0^+)$	$(2^+,-1^-)$ $(2^-,-1^+)$ $(1^+,0^-)$ $(1^-,0^+)$	$(2^-,-1^-)$ $(1^-,0^-)$
0	$(2^+,-2^+)$ $(1^+,-1^+)$	$(2^+,-2^-)$ $(2^-,-2^+)$ $(1^+,-1^-)$ $(1^-,-1^+)$ $(0^+,0^-)$	$(2^-,-2^-)$ $(1^-,-1^-)$

of 21 (7 × 3), and the remaining twenty microstates have combinations of M_L = 3, 2, 1, 0, −1, −2, −3 and M_S = 1, 0, −1 (excluding $2^+,1^+$ already identified). For example, for M_L = 3, M_S = 1, 0, -1 so a microstate from each of the three boxes in the M_L = 3 row need to be removed; the process is then repeated removing a microstate from each of the three boxes in the M_L = 2 row etc. These, and the other microstates of the ^3F term, are shown in Table A.3 in bold e.g. **$(2^+,1^+)$** and should be crossed off the table.

Table A.3 Microstate table for a d^2 configuration grouping microstates of the same term [a]

M_L	M_S 1	0	-1
4		$(2^+,2^-)$	
3	**$(2^+,1^+)$**	$(2^+,1^-)$ **$(2^-,1^+)$**	**$(2^-,1^-)$**
2	**$(2^+,0^+)$**	$(2^+,0^-)$ **$(2^-,0^+)$** *$(1^+,1^-)$*	**$(2^-,0^-)$**
1	**$(2^+,-1^+)$** $(1^+,0^+)$	$(2^+,-1^-)$ **$(2^-,-1^+)$** *$(1^+,0^-)$* $(1^-,0^+)$	**$(2^-,-1^-)$** $(1^-,0^-)$
0	**$(2^+,-2^+)$** $(1^+,-1^+)$	$(2^+,-2^-)$ **$(2^-,-2^+)$** *$(1^+,-1^-)$* $(1^-,-1^+)$ $(0^+,0^-)$	**$(2^-,-2^-)$** $(1^-,-1^-)$
-1	**$(-2^+,1^+)$** $(1^+,0^+)$	$(-2^+,1^-)$ **$(-2^-,1^+)$** *$(-1^+,0^-)$* $(-1^-,0^+)$	**$(-2^-,1^-)$** $(-1^-,0^-)$
-2	**$(-2^+,0^+)$**	$(-2^+,0^-)$ **$(-2^-,0^+)$** *$(-1^+,-1^-)$*	**$(-2^-,0^-)$**
-3	**$(-2^+,-1^+)$**	$(-2^+,-1^-)$ **$(-2^-,-1^+)$**	**$(-2^-,-1^-)$**
-4		$(-2^+,-2^-)$	

[a] ^1G shaded e.g. $(2^+,2^-)$; ^3F in bold e.g. **$(2^+,1^+)$**; ^1D in italic e.g. *$(1^+,1^-)$*; ^3P in underscore e.g. $(1^+,0^+)$; ^1S in standard font e.g. $(0^+,0^-)$.

The process is now continued, starting with $(1^+,1^-)$ which is the microstate with the highest M_L/M_S remaining. This is the start of five microstates belonging to a ^1D term (remove these), then nine microstates forming a ^3P term (remove these), leaving, finally, a single ^1S microstate. This is left as an exercise for the reader.

APPENDIX 3
Answers to SAQs

Chapter 1

SAQ 1.1

There are C_4 and C_2 axes which are coincident and pass through the centre of the ring, perpendicular to the ring plane. Additional C_2 axes are indicated in the figure. The principal axis is C_4, the axis with highest n.

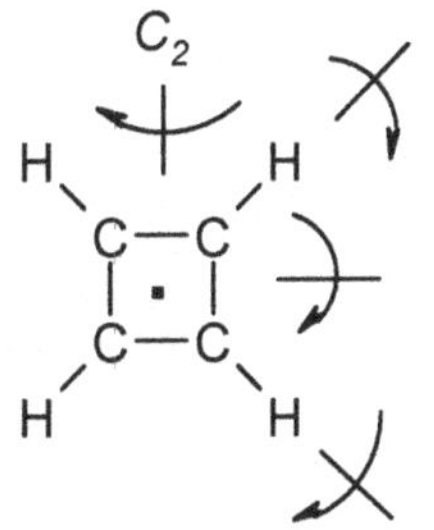

C_2

H H C—C C—C H H

SAQ 1.2

σ_h contains molybdenum and the four CO groups; there are two σ_v planes which contain Mo and pairs of CO groups mutually *trans* to each other, and there are two σ_d which contain molybdenum and bisect the angles C-Mo-C.

SAQ 1.3

In all questions of this type, it is imperative to first determine the correct molecular shape, using VSEPR if the compound has one of the *s*- or *p*-block elements as the central atom. In this case, I = 7e, 7 fluorines contribute a total of 7e, making 14e i.e. 7 e pairs in total. As there are 7 bonds, there are no residual lone pairs and the molecule has a pentagonal bipyramidal shape.

The S_5 axis is coincident with the axial F_1-I-F_2 fragment. S_5^1 moves each of the equatorial fluorines around by one position (a rotation of 72°) but leaves F_1 and F_2 fixed. The reflection part of the operation, which takes place with respect to the equatorial IF_5 plane, then exchanges F_1 and F_2. After S_5^5, the five "rotation through 72° then reflect" operations have brought all the equatorial fluorines back to their original positions, but the odd number of reflections required by S_5^5 means that F_1 and F_2 are still swapped around. It is only when ten "rotation-reflection" operations have been carried out i.e. S_5^{10} that the original position is reverted to. Thus, in general, when n is odd, $S_n^{2n} \equiv E$.

SAQ 1.4

After five S_5 operations all the equatorial fluorines have returned to their original locations, but the odd number of reflections means that the axial fluorines have swapped positions. S_5^5 is thus equivalent to σ_h.

SAQ 1.5

$S_2 \equiv i$. The S_2 point group arises when a molecule possesses only an S_2 axis i.e. S_2 in combination with C_1. Since an S_2 improper axis is equivalent to i ($C_2 + \sigma_h$), then the S_2 point group is equivalent to the C_i point group (only a inversion centre in addition to C_1).

SAQ 1.6

The shape of PF_5 is trigonal bipyramidal. It is not one of the high symmetry linear or cubic point groups though there is a C_3 principal axis (along F_{ax}-P-F_{ax}). There are three C_2 axis perpendicular to C_3, (along each P-F_{eq} bond) and the PF_3 equatorial plane is a σ_h. The point group is therefore D_{3h}.

SAQ 1.7

(a) C_s ; not chiral but polar.
(b) C_2 ; chiral and polar.
(c) D_{2d} ; neither chiral nor polar.

The symmetry elements in (c) and (d) are most easily seen in Newman projections:

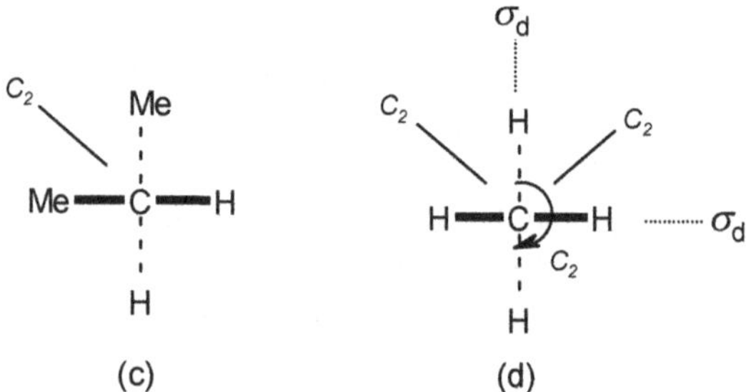

Chapter 2

SAQ 2.1

	E	C_2	$\sigma(xz)$	$\sigma(yz)$
E	E, $\sigma(yz)$	C_2, $\sigma(xz)$	C_2, $\sigma(xz)$	E, $\sigma(yz)$
C_2	C_2, $\sigma(xz)$	E, $\sigma(yz)$	E, $\sigma(yz)$	C_2, $\sigma(xz)$
$\sigma(xz)$	C_2, $\sigma(xz)$	E, $\sigma(yz)$	E, $\sigma(yz)$	C_2, $\sigma(xz)$
$\sigma(yz)$	E, $\sigma(yz)$	C_2, $\sigma(xz)$	C_2, $\sigma(xz)$	E, $\sigma(yz)$

SAQ 2.2

The inverse of C_3^1 (rotate through 120°) is C_3^2 (rotate through 240°)

i.e. $C_3^1 \times C_3^2 = E$.

The inverse of S_5^3 is S_5^7, as $S_5^{10} \equiv E$. Although S_5^2 would complete a 360° rotation, the odd number of reflections requires a second series of five improper rotations (*see SAQ 1.3*), so S_5^2 is not the inverse of S_5^3.

SAQ 2.3

The product matrix has 3 rows and 1 column.

$$\begin{bmatrix} 2 & 4 & -1 \\ 3 & 5 & 0 \\ -2 & 7 & 3 \end{bmatrix}\begin{bmatrix} 2 \\ 3 \\ -2 \end{bmatrix} = \begin{bmatrix} 18 \\ 21 \\ 11 \end{bmatrix} \quad \textit{e.g. } z_{21} = (3 \times 2) + (5 \times 3) + (0 \times -2) = 21$$

SAQ 2.4

$$\begin{bmatrix} 1 & 0 & 0 & 0 & 0 \\ 0 & 0 & 0 & 0 & 1 \\ 0 & 0 & 0 & 1 & 0 \\ 0 & 1 & 0 & 0 & 0 \\ 0 & 0 & 1 & 0 & 0 \end{bmatrix}\begin{bmatrix} C \\ H_1 \\ H_2 \\ H_3 \\ H_4 \end{bmatrix} = \begin{bmatrix} C \\ H_4 \\ H_3 \\ H_1 \\ H_2 \end{bmatrix}$$

SAQ 2.5

$$\begin{aligned} (E)\,\mathbf{T_x} &= (1)\,(\mathbf{T_x}) \\ (C_2)\,\mathbf{T_x} &= (-1)\,(\mathbf{T_x}) \\ (\sigma(xz))\,\mathbf{T_x} &= (1)\,(\mathbf{T_x}) \\ (\sigma(yz))\,\mathbf{T_x} &= (-1)\,(\mathbf{T_x}) \end{aligned}$$

SAQ 2.6

$[\sigma(xz) \times \sigma(yz)] \times C_2 = [(1) \times (-1)] \times (-1) = 1$

$\sigma(xz) \times [\sigma(yz) \times C_2] = (1) \times [(-1) \times (-1)] = 1$

SAQ 2.7

For example:

$\sigma(xz) \times \sigma(yz) = C_2$, but $-1 \times -1 \neq -1$

SAQ 2.8

$$C_3^{\,1} \times C_3^{\,2} = E$$

For the representations based on $\mathbf{T_z}$ and $\mathbf{R_z}$, this becomes 1×1 which = 1, the character for E in each case.

For the representations based on ($\mathbf{T_x}$, $\mathbf{T_z}$) and ($\mathbf{R_x}$, $\mathbf{R_z}$):

$$\begin{bmatrix} -\tfrac{1}{2} & -\tfrac{\sqrt{3}}{2} \\ \tfrac{\sqrt{3}}{2} & -\tfrac{1}{2} \end{bmatrix}\begin{bmatrix} -\tfrac{1}{2} & \tfrac{\sqrt{3}}{2} \\ -\tfrac{\sqrt{3}}{2} & -\tfrac{1}{2} \end{bmatrix} =$$

$$\begin{bmatrix} \left(-\tfrac{1}{2} \times -\tfrac{1}{2}\right)+\left(-\tfrac{\sqrt{3}}{2} \times -\tfrac{\sqrt{3}}{2}\right) & \left(-\tfrac{1}{2} \times \tfrac{\sqrt{3}}{2}\right)+\left(-\tfrac{\sqrt{3}}{2} \times -\tfrac{1}{2}\right) \\ \left(\tfrac{\sqrt{3}}{2} \times -\tfrac{1}{2}\right)+\left(-\tfrac{1}{2} \times -\tfrac{\sqrt{3}}{2}\right) & \left(\tfrac{\sqrt{3}}{2} \times \tfrac{\sqrt{3}}{2}\right)+\left(-\tfrac{1}{2} \times -\tfrac{1}{2}\right) \end{bmatrix} = \begin{bmatrix} 1 & 0 \\ 0 & 1 \end{bmatrix}$$

SAQ 2.9

Matrix	χ	Matrix	χ
$\begin{bmatrix} 1 & 0 \\ 0 & 1 \end{bmatrix}$	$1+1=2$	$\begin{bmatrix} 1 & 0 \\ 0 & -1 \end{bmatrix}$	$1+(-1)=0$
$\begin{bmatrix} -\frac{1}{2} & -\frac{\sqrt{3}}{2} \\ \frac{\sqrt{3}}{2} & -\frac{1}{2} \end{bmatrix}$	$-\frac{1}{2}+(-\frac{1}{2})=-1$	$\begin{bmatrix} -\frac{1}{2} & -\frac{\sqrt{3}}{2} \\ -\frac{\sqrt{3}}{2} & \frac{1}{2} \end{bmatrix}$	$-\frac{1}{2}+\frac{1}{2}=0$
$\begin{bmatrix} -\frac{1}{2} & \frac{\sqrt{3}}{2} \\ -\frac{\sqrt{3}}{2} & -\frac{1}{2} \end{bmatrix}$	$-\frac{1}{2}+(-\frac{1}{2})=-1$	$\begin{bmatrix} -\frac{1}{2} & \frac{\sqrt{3}}{2} \\ \frac{\sqrt{3}}{2} & \frac{1}{2} \end{bmatrix}$	$-\frac{1}{2}+\frac{1}{2}=0$

Chapter 3

SAQ 3.1

$$\begin{bmatrix} -1 & 0 & 0 & 0 & 0 & 0 & 0 & 0 & 0 \\ 0 & 1 & 0 & 0 & 0 & 0 & 0 & 0 & 0 \\ 0 & 0 & 1 & 0 & 0 & 0 & 0 & 0 & 0 \\ 0 & 0 & 0 & -1 & 0 & 0 & 0 & 0 & 0 \\ 0 & 0 & 0 & 0 & 1 & 0 & 0 & 0 & 0 \\ 0 & 0 & 0 & 0 & 0 & 1 & 0 & 0 & 0 \\ 0 & 0 & 0 & 0 & 0 & 0 & -1 & 0 & 0 \\ 0 & 0 & 0 & 0 & 0 & 0 & 0 & 1 & 0 \\ 0 & 0 & 0 & 0 & 0 & 0 & 0 & 0 & 1 \end{bmatrix} \begin{bmatrix} S_x \\ S_y \\ S_z \\ O^1_x \\ O^1_y \\ O^1_z \\ O^2_x \\ O^2_y \\ O^2_z \end{bmatrix} = \begin{bmatrix} -S_x \\ S_y \\ S_z \\ -O^1_x \\ O^1_y \\ O^1_z \\ -O^2_x \\ O^2_y \\ O^2_z \end{bmatrix}$$

All y, z vectors are unmoved while vectors along x on all three atoms are reversed.

SAQ 3.2

The six vectors lying in the yz plane are unmoved (6×1) while all the vectors lying along x are reversed on reflection in the yz mirror plane (3×-1).

Thus, $\chi = 6 - 3 = 3$.

SAQ 3.3

Total number of operations (g) = 6 (E, $2C_3$, $3\sigma_v$). Remember, the "$2C_3$" refers to the two operations C_3^1 and C_3^2 and not two C_3 axes. C_3^3 is already in the character table as E.

The number of operations in each class (n_R) is 1 (E), 2 (C_3) and 3 ($3\sigma_v$).

SAQ 3.4

C_{4v}	E	$2C_4$	C_2	$2\sigma_v$	$2\sigma_d$
Γ_{3N}	21	3	–3	5	3

E : 21 × 1, as all vectors unmoved.

C_4 : Only the vectors on W and the halogens on the z-axis are unmoved, all other vectors move. For the three atoms on z, their vector on z is unmoved (3 × 1) while their vectors along x and y rotate by 90° to new positions (6 × 0).

C_2 : Only vectors associated with the atoms lying on the z-axis contribute (others: 12 × 0); the vectors along z for these atoms are unmoved (3 × 1), though their x, y vectors rotate through 180^0 to their reverse (6 × –1).

σ_v : Only the five atoms lying on the xz mirror plane contribute. For each atom, two vectors lie in the xz mirror plane (10 × 1), while the y vector on each atom is reversed (5 × –1).

σ_d : Only the vectors on the three atoms on z contribute; the x, y vectors swap places (6 × 0) while their z vectors are unmoved (3 × 1).

$$A_1 = 1/8\,[(1\times 21\times 1)+(2\times 3\times 1)+(1\times -3\times 1)+(2\times 5\times 1)+(2\times 3\times 1)] = 5$$
$$A_2 = 1/8\,[(1\times 21\times 1)+(2\times 3\times 1)+(1\times -3\times 1)+(2\times 5\times -1)+(2\times 3\times -1)] = 1$$
$$B_1 = 1/8\,[(1\times 21\times 1)+(2\times 3\times -1)+(1\times -3\times 1)+(2\times 5\times 1)+(2\times 3\times -1)] = 2$$
$$B_2 = 1/8\,[(1\times 21\times 1)+(2\times 3\times -1)+(1\times -3\times 1)+(2\times 5\times -1)+(2\times 3\times 1)] = 1$$
$$E = 1/8\,[(1\times 21\times 2)+(2\times 3\times 0)+(1\times -3\times -2)+(2\times 5\times 0)+(2\times 3\times 0)] = 6$$

$$\Gamma_{3N} = 5A_1 + A_2 + 2B_1 + B_2 + 6E \qquad (3N = 21)$$

$$\Gamma_{trans+rot} = A_1 + A_2 + 2E$$

We expect three translations and three rotations, and we have labels describing these six movements since each E describes a doubly degenerate movement ($\mathbf{T_x}$, $\mathbf{T_y}$; $\mathbf{R_x}$, $\mathbf{R_y}$) (Section 2.5).

$$\Gamma_{vib} = \Gamma_{3N} - \Gamma_{trans+rot} = 4A_1 + 2B_1 + B_2 + 4E$$

This totals 15 vibrational modes consistent with 3N-6 (note: again each E is a doubly degenerate mode).

SAQ 3.5

For S_4, $\chi_{u.a.} = -1 + 2\cos\theta = -1 + 2\cos(90) = -1 + 2(0) = -1$.

SAQ 3.6

$\chi_{u.a.} = 1 + 2\cos\theta$

For C_2 : $\chi_{u.a.} = 1 + 2\cos(180) = 1 + 2(-1) = -1$

For C_4 : $\chi_{u.a.} = 1 + 2\cos(90) = 1 + 2(0) = 1$

Other $\chi_{u.a.}$ are given in Section 3.4: E = 3, σ = 1

C_{4v}	E	$2C_4$	C_2	$2\sigma_v$	$2\sigma_d$
unshifted atoms	7	3	3	5	3
× $\chi_{u.a.}$	× 3	× 1	× –1	× 1	× 1
Γ_{3N}	21	3	–3	5	3

(as in SAQ 3.4)

Chapter 4

SAQ 4.1

Only A_{2u} (same symmetry as $\mathbf{T_z}$) and E_u (same symmetry as $\mathbf{T_x}$, $\mathbf{T_y}$) are infrared active; all the other modes are infrared inactive.

SAQ 4.2

A_{1g} ($x^2 + y^2$, z^2), B_{1g} ($x^2 - y^2$) and B_{2g} (xy) are all Raman active, while the remaining modes are Raman inactive.

SAQ 4.3

FN=NF can exist as either *cis-* or *trans-* isomers:

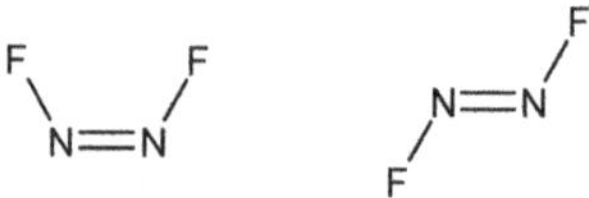

As there are no coincidences in the infrared and Raman spectra, the molecule must possess an inversion centre and is thus the *trans*-isomer; the inversion centre is at the mid-point of the N-N bond.

Chapter 5

SAQ 5.1

D_{4h}	E	$2C_4$	C_2	$2C_2'$	$2C_2''$	i	$2S_4$	σ_h	$2\sigma_v$	$2\sigma_d$
un.atoms	5	1	1	3	1	1	1	5	3	1
$\times \chi_{u.a.}$	3	1	−1	−1	−1	−3	−1	1	1	1
Γ_{3N}	15	1	−1	−3	−1	−3	−1	5	3	1

Using the reduction formula:

$$\Gamma_{3N} = A_{1g} + A_{2g} + B_{1g} + B_{2g} + E_g + 2A_{2u} + B_{2u} + 3E_u$$

$$\Gamma_{trans+rot} = A_{2g} + E_g + A_{2u} + E_u$$

$$\Gamma_{vib} = \Gamma_{3N} - \Gamma_{trans+rot}$$

SAQ 5.2

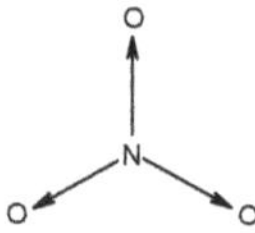

The nitrate ion belongs to the D_{3h} point group, assuming complete delocalisation of the negative charge, making all three N-O bonds equivalent.

D_{3h}	E	$2C_3$	$3C_2$	σ_h	$2S_3$	$3\sigma_v$
$\Gamma_{N\text{-}O}$	3	0	1	3	0	1

$= A_1' + E'$

A_1' : Raman-only (pol) active.
E' : Infrared and Raman (depol) active.

SAQ 5.3

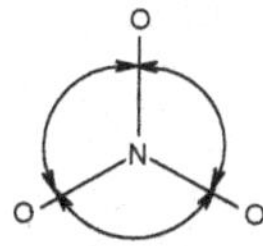

D_{3h}	E	$2C_3$	$3C_2$	σ_h	$2S_3$	$3\sigma_v$
$\Gamma_{\text{in-plane}}$	3	0	1	3	0	1

= A_1' + E', as above

Note that under both C_2 and σ_v one double-headed arrow swaps ends but remains indistinguishable from the original (count 1) while two double-headed arrows move to new locations (count 0, 0).

The A_1' mode is a redundancy, as it has already been accounted for by an N-O stretching mode (*see SAQ 5.2*). As with the example of $[PtCl_4]^{2-}$, this corresponds to all bond angles opening or closing simultaneously.

SAQ 5.4

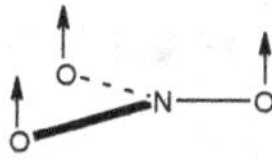

D_{3h}	E	$2C_3$	$3C_2$	σ_h	$2S_3$	$3\sigma_v$
$\Gamma_{\text{out-of-plane}}$	3	0	−1	−3	0	1

= A_2'' + E''

Given:

$$\Gamma_{\text{vib}} = A_1' + 2E' + A_2''$$
$$\Gamma_{\text{N-O}} = A_1' + E'$$
$$\Gamma_{\text{in-plane}} = E'$$

you should predict:

$$\Gamma_{\text{out-of-plane}} = A_2''$$

The E'' is a redundancy and corresponds to rotation of the whole molecule about x, y ($\mathbf{R}_x$, $\mathbf{R}_y$).

SAQ 5.5

$$\Gamma_{\text{vib}} = A_1' + 2E' + A_2''$$

Infrared (solid)	Raman (solution)		
1383	1385 (depol)	E'	N-O stretch
	1048 (pol)	A_1'	N-O stretch
825		A_2"	out-of-plane bend
720	718 (depol)	E'	in-plane bend

The A_1' band is the Raman-only polarised band and is an N-O stretch.
The A_2" band is the infrared-only band and is an out-of-plane bend.
The two E' modes are active in both the infrared and Raman (depol); they can be distinguished by the N-O stretch coming at higher energy than the in-plane bend.

Chapter 6

SAQ 6.1

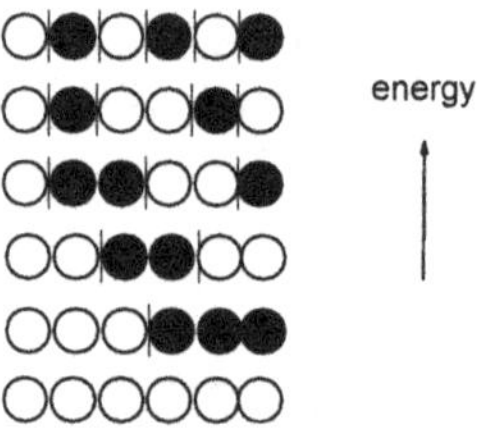

SAQ 6.2

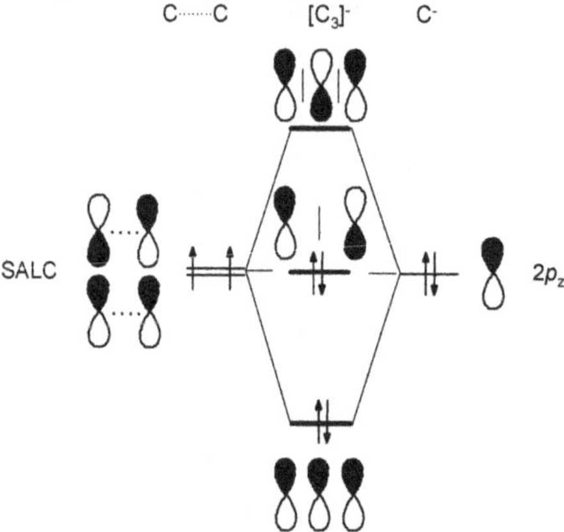

The three p_z orbitals combine to form three π-MOs (assuming the molecule lies in the *xy* plane). There are two ways in which the terminal atom p_z AOs can combine (*left*) and only the in-phase combination can combine with the p_z on the central carbon, to generate bonding and anti-bonding MOs. The out-of-phase combination of terminal atom p_z AOs is non-bonding.

Each carbon has four electrons, three of which are in sp^2 hybrids, leaving p_z with 1 electron for π-bonding. Along with the electron for the negative charge on the anion, this gives four π-electrons which fill the bonding and non-bonding MOs. The π-bond order is 1 i.e. 0.5 per C-C bond. This correlates with the following resonance forms of the allyl anion:

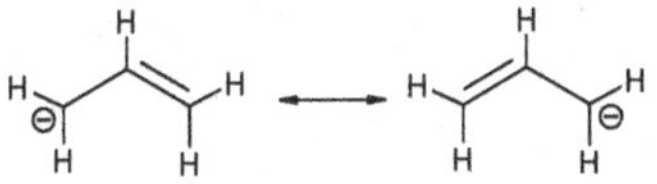

Chapter 7

SAQ 7.1

s : a_{1g} (always transforms as the perfectly symmetrical representation)
p_x, p_y : e_u (same as $\mathbf{T_x}$, $\mathbf{T_y}$)
p_z : a_{2u} (same as $\mathbf{T_z}$)

Chapter 8

SAQ 8.1

C_{3v}	E	$2C_3$	$3\sigma_v$
$\Gamma_{H\,1s}$	3	0	1

$= a_1 + e$

This is reasonable as three combinations are expected; one SALC is unique (a_1) while the other two form a degenerate pair (e).

Symmetries of AOs on nitrogen:

$2s$: a_1 (all s-orbitals are perfectly symmetrical)
$2p_x, 2p_y$: e ($\mathbf{T_x}$, $\mathbf{T_y}$)
$2p_z$: a_1 ($\mathbf{T_z}$)

SALC	AO	Label	MOs
	$2s$	a_1	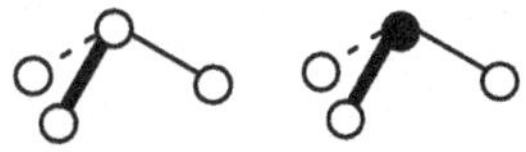
	$2p_x$	e	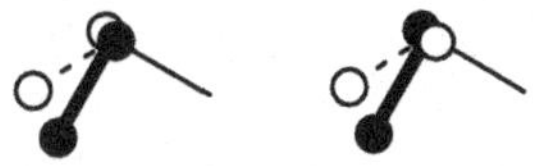
	$2p_y$	e	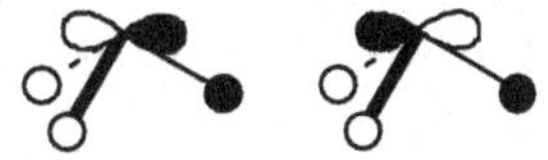
	$2p_z$	a_1	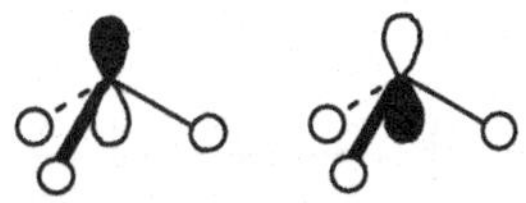

SAQ 8.2

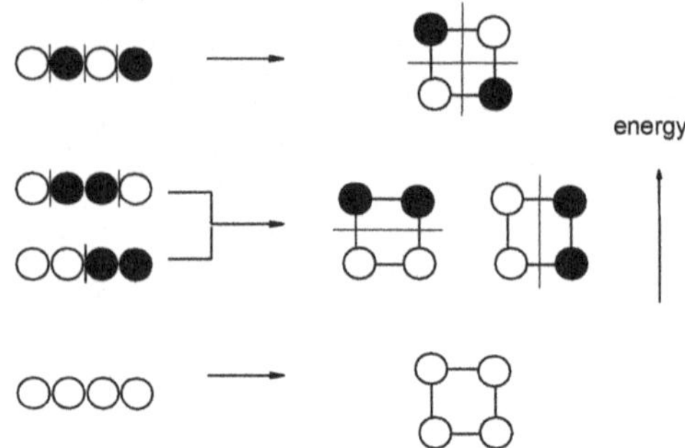

Chapter 9

SAQ 9.1

O_h	E	$8C_3$	$6C_2$	$6C_4$	$3C_2$[a]	i	$6S_4$	$8S_6$	$3\sigma_h$	$6\sigma_d$
$\Gamma_{d\text{-orbitals}}$	5	−1	1	−1	1	5	−1	−1	1	1

[a] $=C_4^2$

This can be converted to e_g and t_{2g} using the reduction formula, or:

O_h	E	$8C_3$	$6C_2$	$6C_4$	$3C_2$	i	$6S_4$	$8S_6$	$3\sigma_h$	$6\sigma_d$
E_g	2	−1	0	0	2	2	0	−1	2	0
T_{2g}	3	0	1	−1	−1	3	−1	0	−1	1
$\Gamma_{d\text{-orbitals}}$	5	−1	1	−1	1	5	−1	−1	1	1

SAQ 9.2

O_h		D_{4h}	
d_{xy}, d_{xz}, d_{yz}	: t_{2g}	d_{xy}	: b_{2g}
		d_{xz}, d_{yz}	: e_g
$d_{z^2}, d_{x^2-y^2}$	: e_g	$d_{x^2-y^2}$	: b_{1g}
		d_{z^2}	: a_{1g}

As the ligand along z is removed the e_g degeneracy is lost. Similarly, the t_{2g} trio are separated, with d_{xy} differing from the two d-orbitals with a z component.

SAQ 9.3

T_d	E	$8C_3$	$3C_2$	$6S_4$	$6\sigma_d$	
Γ_{4L}	4	1	0	0	2	$= a_1 + t_2$

The AO symmetries are :

s	:	a_1
p_x, p_y, p_z	:	t_2
$d_{z^2}, d_{x^2-y^2}$	:	e
d_{xy}, d_{xz}, d_{yz}	:	t_2

Remembering that the AO energies are $d < s < p$, a qualitative MO diagram is:

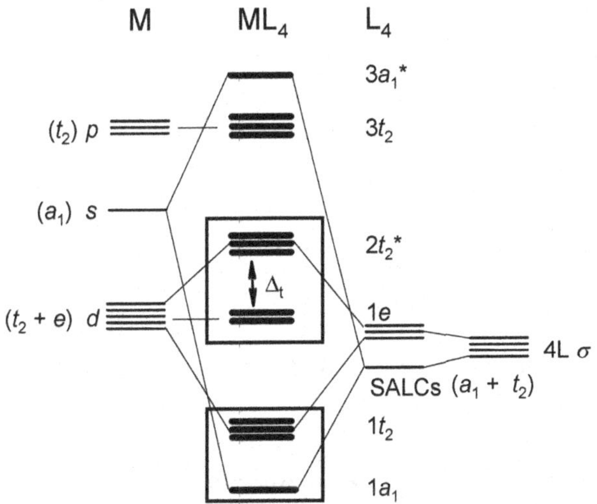

The 8 ligand electrons (4 × 2) fill the $1a_1$ and $1t_2$ MOs, while the metal-based M^{n+} electrons fill the $1e$ and $2t_2$*. The $1e$ - $2t_2$* energy gap corresponds to Δ_t in crystal field theory, though here it is described as an e - t_2 separation.

Note that the t_2 AOs p_x, p_y, p_z could interact with the other orbitals of t_2 symmetry (particularly the SALCs) if the energy match is correct. This would lower $1t_2$ and $2t_2$* while raising the energy of $3t_2$. $2t_2$* would become more non-bonding in character while $3t_2$ would become anti-bonding. However, the overall result of $1a_1$ and $1t_2$ accommodating the ligand electrons and $1e$, $2t_2^*$ the metal electrons remains unchanged.

Chapter 11

SAQ 11.1

C_{2V}	E	C_2	$\sigma(xz)$	$\sigma(yz)$	
B_1	1	−1	1	−1	
B_1	1	−1	1	−1	
$B_1 \times B_1$	1	1	1	1	= A_1

SAQ 11.2

C_{2v}	E	C_2	$\sigma(xz)$	$\sigma(yz)$
B_1	1	−1	1	−1
A_1	1	1	1	1
$A_1 \times A_1 \times B_1 \times A_1$	1	−1	1	−1

$= B_1$

SAQ 11.3

$$A_1 \times \begin{pmatrix} \mu_x \\ \mu_y \\ \mu_z \end{pmatrix} \times A_2 \quad = \quad A_1 \times \begin{pmatrix} \mathbf{T}_x \\ \mathbf{T}_y \\ \mathbf{T}_z \end{pmatrix} \times A_2 \quad = \quad A_1 \times \begin{pmatrix} B_1 \\ B_2 \\ A_1 \end{pmatrix} \times A_2$$

The direct products are:

$$A_1 \times B_1 \times A_2 = B_2$$
$$A_1 \times B_2 \times A_2 = B_1$$
$$A_1 \times A_1 \times A_2 = A_2$$

As none of these direct products are A_1 the transition is forbidden.

SAQ 11.4

The allowed transition (a) is $^1A_1 \rightarrow {}^1B_1$, while (b) is $^1A_1 \rightarrow {}^3B_1$ and spin forbidden.

SAQ 11.5

C_{6v}	E	$2C_6$	$2C_3$	C_2	$3\sigma_v$	$3\sigma_d$
E_1	2	1	−1	−2	0	0
E_2	2	−1	−1	2	0	0
$E_1 \times E_2$	4	−1	1	−4	0	0

$= B_1 + B_2 + E_1$

SAQ 11.6

$C_{2h} : (A_g \times A_u) \times B_u = A_u \times B_u = B_g$
$D_{3h} : (A_1'' \times E'') \times A_2' = E' \times A_2' = E'$
$T_d : (E \times T_1) \times T_2 = (T_1 + T_2) \times T_2 = (T_1 \times T_2) + (T_2 \times T_2)$
$= A_1 + A_2 + 2E + 2T_1 + 2T_2$
$O_h : (E_g \times A_{2g}) \times T_{1u} = E_g \times T_{1u} = T_{1u} + T_{2u}$

SAQ 11.7

Under O_h symmetry, μ has T_{1u} symmetry.

$A_{1u} \rightarrow T_{2g} : A_{1u} \times T_{2g} = T_{2u}$. This does not match T_{1u} so the transition is forbidden.

$E_u \rightarrow T_{1g}$: $E_u \times T_{1g} = T_{1u} + T_{2u}$. As this contains T_{1u} so the transition is allowed.

Chapter 12

SAQ 12.1

$D_t = (2 \times 5)! / 2! \times 8! \quad = \quad 45$

SAQ 12.2

C_{2v}	E	C_2	$\sigma(xz)$	$\sigma(yz)$
$\Gamma_{p\text{-orbitals}}$	3	−1	1	1

$= A_1 + B_1 + B_2$

These correspond to $\mathbf{T_z}$, $\mathbf{T_x}$, $\mathbf{T_y}$, repectively, as expected.

SAQ 12.3

$E_g \times E_g = A_{1g} + A_{2g} + E_g$

For $(e_g)^2$: $D_t = (2 \times 2)! / 2! \times 2! = 6$

SAQ 12.4

^4F has a maximum $M_L = 3$ ($m_l = 2 + 1 + 0$). Thus, $M_L = 3, 2, 1, 0, -1, -2, -3$ and $S = {}^3/_2$ (3 x $^1/_2$) so $M_S = {}^3/_2, {}^1/_2, -{}^1/_2, -{}^3/_2$. The total mumber of microstates is $7 \times 4 = 28$. This can also be arrived at by $(2S + 1)(2L + 1)$.

In a weak octahedral field, ^4F spilts into $^4A_{2g}$ + $^4T_{1g}$ + $^4T_{2g}$ each of which corresponds to 4, 12 and 12 microstates, respectively.

SAQ 12.5

As the ^{1}G term arises from a combination of *d*-orbitals the terms must have *g* symmetry and "+" should be used in the relevant equations.

O_h	E	$8C_3$	$6C_2$	$6C_4$	$3C_2{}^a$	i	$6S_4$	$8S_6$	$3\sigma_h$	$6\sigma_d$
Γ_G	9	0	1	1	1	9	1	0	1	1

$= A_{1g} + E_g + T_{1g} + T_{2g}$

The singlet nature of the ^{1}G term carries through to the ligand-field terms, so these become $^1A_{1g}$ + ^1E_g + $^1T_{1g}$ + $^1T_{2g}$. They total 1 + 2 + 3 + 3 = 9 microstates, as required (*Section 12.1*).

SAQ 12.6

High-spin $d^7 = (t_{2g})^5(e_g)^2 = {}^2T_{2g} \times {}^3A_{2g} = {}^4T_{1g}$

Low-spin $d^7 = (t_{2g})^6(e_g)^1 = {}^1A_{1g} \times {}^2E_g = {}^2E_g$

Chapter 13

SAQ 13.1

O_h	E	$8C_3$	$6C_2$	$6C_4$	$3C_2^a$	i	$6S_4$	$8S_6$	$3\sigma_h$	$6\sigma_d$	
A_{1g}	1	1	1	1	1	1	1	1	1	1	
T_{1u}	3	0	−1	1	−1	−3	−1	0	0	1	
$A_{1g} \times T_{1u} =$	3	0	−1	1	−1	−3	−1	0	0	1	$= T_{1u}$

SAQ 13. 2

$\Gamma_{vib} = A_{1g} + E_g + 2T_{1u} + T_{2g} + T_{2u}$

For $^1T_{1g} \leftarrow {}^1A_{1g}$:

$$\begin{aligned} T_{1g} \times T_{1u} \times A_{1g} &= (A_{1u} + E_u + T_{1u} + T_{2u}) \times A_{1g} \\ &= A_{1u} + E_u + T_{1u} + T_{2u} \end{aligned}$$

As this spans Γ_{vib} (T_{1u}, T_{2u} are common to both), the transition is vibronically allowed.

For $^1T_{2g} \leftarrow {}^1A_{1g}$:

$$\begin{aligned} T_{2g} \times T_{1u} \times A_{1g} &= (A_{2u} + E_u + T_{1u} + T_{2u}) \times A_{1g} \\ &= A_{2u} + E_u + T_{1u} + T_{2u} \end{aligned}$$

which is also vibronically-allowed.

SAQ 13.3

High-spin Co^{3+} is d^6, and the spin allowed transition is $^5E_g \leftarrow {}^5T_{2g}$

SAQ 13.4

$[V(H_2O)_6]^{3+}$ is V^{3+} and d^2 so we expect $^3T_{2g} \leftarrow {}^3T_{1g}(F)$, $^3T_{1g}(P) \leftarrow {}^3T_{1g}(F)$ and $^3A_{2g} \leftarrow {}^3T_{1g}(F)$ spin-allowed transitions.

SAQ 13.5

$[FeCl_4]^{2-}$ is d^6 ($\equiv$ octahedral d^4) so a single $^5T_2 \leftarrow {}^5E$ transition would be expected, and one is seen at *ca.* 4,000 cm^{-1}.

SAQ 13.6

$[NiCl_4]^{2-}$ is d^8 ($\equiv$ octahedral d^2) so the three spin-allowed transitions are :

$^3T_2 \leftarrow {}^3T_1(F)$, $^3T_1(P) \leftarrow {}^3T_1(F)$ and $^3A_2 \leftarrow {}^3T_1(F)$

SAQ 13.7

The three possible transitions are $^3T_2 \leftarrow {}^3A_2$, $^3T_1(F) \leftarrow {}^3A_2$ and $^3T_1(P) \leftarrow {}^3A_2$

The symmetry of the dipole moment is T_2 under T_d symmetry, so the transition integrals are:

$$A_2 \times T_2 \times T_2 = T_1 \times T_2 = A_2 + E + T_1 + T_2$$
$$A_2 \times T_2 \times T_1 = T_1 \times T_1 = A_1 + E + T_1 + T_2 \text{ (for two transitions)}$$

The $^3T_2 \leftarrow {}^3A_2$ excitation is symmetry-forbidden while the two $^3T_1 \leftarrow {}^3A_2$ transitions are allowed as they span the totally symmetric representation A_1.

SAQ 13.8

$(t_{2g})^6$ has $^1A_{1g}$ symmetry, therefore singlet states associated with $(t_{2g})^5(e_g)^1$ need to be identified.

$$(t_{2g})^5(e_g)^1 \equiv (t_{2g})^1(e_g)^1 \text{ (by the hole formalism)} = {}^2T_{2g} \times {}^2E_g$$
$$^2T_{2g} \times {}^2E_g = {}^1T_{1g} + {}^1T_{2g} + {}^3T_{1g} + {}^3T_{2g}$$

Spin-allowed transitions:

$^1T_{1g} \leftarrow {}^1A_{1g}$ and $^1T_{2g} \leftarrow {}^1A_{1g}$

SAQ 13.9

	E	$6C_4$	$3C_2$[a]		$6C_2$	i	$6S_4$	$3\sigma_h$		$6\sigma_d$	O_h
D_{4h}	E	$2C_4$	C_2	$2C_2'$	$2C_2''$	i	$2S_4$	σ_h	$2\sigma_v$	$2\sigma_d$	
B_{2u}	1	−1	1	−1	1	−1	1	−1	1	−1	
E_u	2	0	−2	0	0	−2	0	2	0	0	
$B_{2u} + E_u$	3	−1	−1	−1	1	−3	1	1	1	−1	T_{2u}

[a] $= C_4^2$

SAQ 13.10

$[Cr(H_2O)_6]^{2+}$ is Cr(II) d^4 $[(t_{2g})^3(e_g)^1]$ and is subject to a Jahn-Teller distortion. The ^5E_g ground term splits into $^5B_{1g}$ and $^5A_{1g}$, while the $^5T_{2g}$ exhibits a much smaller splitting into $^5B_{2g}$ and ^5E_g, analogous to d^9. The spectrum can be interpreted in terms of $^5A_{1g} \leftarrow {}^5B_{1g}$ and unresolved $^5B_{2g} \leftarrow {}^5B_{1g}$ / $^5E_g \leftarrow {}^5B_{1g}$ transitions.

Appendix 1

SAQ A.1

C_{2v}	E	C_2	$\sigma(xz)$	$\sigma(yz)$	
$\psi_1 \rightarrow$	ψ_1	ψ_2	ψ_2	ψ_1	
$\chi_{(IR)}\, b_2$	1	−1	−1	1	
$\chi_{(IR)}R\psi_1$	ψ_1	$-\psi_2$	$-\psi_2$	ψ_1	$= 2\psi_1 - 2\psi_2$

Since we are only interested in the relative contributions of the AOs, this is the same as $\psi_1 - \psi_2$. Including the normalisation coefficient $b_2 = 1/\sqrt{2}(\psi_1 - \psi_2)$.

SAQ A.2

Multiplying pairs of coefficients ($a_2'' \times e''$) for ψ_1, ψ_2 and ψ_3:

$$(1 \times 2) + (1 \times -1) + (1 \times -1) = 0\text{, so orthogonal}$$

SAQ A.3

D_{3h}	E	C_3^1	C_3^2	σ_h	S_3^1	S_3^5
$\psi_3 \rightarrow$	ψ_3	ψ_1	ψ_2	$-\psi_3$	$-\psi_1$	$-\psi_2$
$\chi_{(IR)}\, e''$	2	-1	-1	-2	1	1
$\chi_{(IR)}R\psi_1$	$2\psi_3$	$-\psi_1$	$-\psi_2$	$2\psi_3$	$-\psi_1$	$-\psi_2$

$e'' = 4\psi_3 - 2\psi_1 - 2\psi_2$ or $e'' = 2\psi_3 - \psi_1 - \psi_2$

$(2 \times -1) + (-1 \times -1) + (-1 \times 2) = -3$, not 0 as required for orthogonality.

SAQ A.4

C_{3v}	E	C_3^1	C_3^2	$\sigma_v(1)$	$\sigma_v(2)$	$\sigma_v(3)$	
$\alpha_1 \rightarrow$	α_1	α_2	α_3	α_3	α_2	α_1	
$\chi_{(IR)}\, A_1$	1	1	1	1	1	1	
$\chi_{(IR)}Rv_1$	α_1	α_2	α_3	α_3	α_2	α_1	$\rightarrow 1/\sqrt{3}(\alpha_1 + \alpha_2 + \alpha_3)$
$\chi_{(IR)}\, E$	2	-1	-1	0	0	0	
$\chi_{(IR)}Rv_1$	$2\alpha_1$	$-\alpha_2$	$-\alpha_3$				$\rightarrow 1/\sqrt{6}(2\alpha_1 - \alpha_2 - \alpha_3)$

C_{3v}	E	C_3^1	C_3^2	
$\alpha_2 - \alpha_3 \rightarrow$	$\alpha_2 - \alpha_3$	$\alpha_3 - \alpha_1$	$\alpha_1 - \alpha_2$	
$\chi_{(IR)}\, E$	2	-1	-1	
$\chi_{(IR)}Rv_1$	$2\alpha_2 - 2\alpha_3$	$-\alpha_3 + \alpha_1$	$-\alpha_1 + \alpha_2$	$\rightarrow 1/\sqrt{2}(\alpha_2 - \alpha_3)$

SAQ A.5

For the singly-degenerate A_{1g} and B_{1g} modes:

D_{4h}	E	C_4^1	C_4^3	C_2	$C_2'(1)$	$C_2'(2)$	$C_2''(1)$	$C_2''(2)$	i	S_4^1
$v_1 \rightarrow$	v_1	v_2	v_4	v_3	v_1	v_3	v_2	v_4	v_3	v_2
$\chi_{(IR)}\, A_{1g}$	1	1	1	1	1	1	1	1	1	1
$\chi_{(IR)}Rv_1$	v_1	v_2	v_4	v_3	v_1	v_3	v_2	v_4	v_3	v_2
$\chi_{(IR)}\, B_{1g}$	1	-1	-1	1	1	1	-1	-1	1	-1
$\chi_{(IR)}Rv_1$	v_1	$-v_2$	$-v_4$	v_3	v_1	v_3	$-v_2$	$-v_4$	v_3	$-v_2$

continued

D_{4h}	S_4^3	σ_h	$\sigma_v(1)$	$\sigma_v(2)$	$\sigma_d(1)$	$\sigma_d(2)$	
$v_1 \rightarrow$	v_4	v_1	v_1	v_3	v_2	v_4	
$\chi_{(IR)}\,A_{1g}$	1	1	1	1	1	1	
$\chi_{(IR)}Rv_1$	v_4	v_1	v_1	v_3	v_2	v_4	$= 4v_1 + 4v_2 + 4v_3 + 4v_4$
$\chi_{(IR)}\,B_{1g}$	-1	1	1	1	-1	-1	
$\chi_{(IR)}Rv_1$	$-v_4$	v_1	v_1	v_3	$-v_2$	$-v_4$	$= 4v_1 - 4v_2 + 4v_3 - 4v_4$

For the doubly-degenerate E_u mode:

D_{4h}	E	C_2	i	σ_h	
$v_1 \rightarrow$	v_1	v_3	v_3	v_1	
$\chi_{(IR)}E_u$	2	-2	-2	2	
$\chi_{(IR)}Rv_1$	$2v_1$	$-2v_3$	$-2v_3$	$2v_1$	$= 4v_1 - 4v_3$
$v_2 \rightarrow$	v_2	v_4	v_4	v_2	
$\chi_{(IR)}E_u$	2	-2	-2	2	
$\chi_{(IR)}Rv_1$	$2v_2$	$-2v_4$	$-2v_4$	$2v_2$	$= 4v_2 - 4v_4$

From which:

$A_{1g} = 1/2(v_1 + v_2 + v_3 + v_4)$

$B_{1g} = 1/2(v_1 - v_2 + v_3 - v_4)$

$E_u = 1/\sqrt{2}(v_1 - v_3),\ 1/\sqrt{2}(v_2 - v_4)$

SAQ A.6

For the singly-degenerate A_{2u} and B_{2u} modes:

D_{4h}	E	C_4^1	C_4^3	C_2	$C_2'(1)$	$C_2'(2)$	$C_2''(1)$	$C_2''(2)$	i	S_4^1
$\beta_1 \rightarrow$	β_1	β_2	β_4	β_3	$-\beta_1$	$-\beta_3$	$-\beta_2$	$-\beta_4$	$-\beta_3$	$-\beta_2$
$\chi_{(IR)}\,A_{2u}$	1	1	1	1	-1	-1	-1	-1	-1	-1
$\chi_{(IR)}Rv_1$	β_1	β_2	β_4	β_3	β_1	β_3	β_2	β_4	β_3	β_2
$\chi_{(IR)}\,B_{2u}$	1	-1	-1	1	-1	-1	1	1	-1	1
$\chi_{(IR)}Rv_1$	β_1	$-\beta_2$	$-\beta_4$	β_3	β_1	β_3	$-\beta_2$	$-\beta_4$	β_3	$-\beta_2$

continued

D_{4h}	S_4^3	σ_h	$\sigma_v(1)$	$\sigma_v(2)$	$\sigma_d(1)$	$\sigma_d(2)$	
$\beta_1 \rightarrow$	$-\beta_4$	$-\beta_1$	β_1	β_3	β_2	β_4	
$\chi_{(IR)}\,A_{2u}$	-1	-1	1	1	1	1	
$\chi_{(IR)}Rv_1$	β_4	β_1	β_1	β_3	β_2	β_4	$= 4\beta_1 + 4\beta_2 + 4\beta_3 + 4\beta_4$
$\chi_{(IR)}\,B_{2u}$	1	-1	1	1	-1	-1	
$\chi_{(IR)}Rv_1$	$-\beta_4$	β_1	β_1	β_3	$-\beta_2$	$-\beta_4$	$= 4\beta_1 - 4\beta_2 + 4\beta_3 - 4\beta_4$

For the doubly-degenerate E_g mode:

D_{4h}	E	C_2	i	σ_h	
$\beta_1 \rightarrow$	β_1	β_3	$-\beta_3$	$-\beta_1$	
$\chi_{(IR)}\,E_g$	2	−2	2	−2	
$\chi_{(IR)}Rv_1$	$2\beta_1$	$-2\beta_3$	$-2\beta_3$	$2\beta_1$	$= 4\beta_1 - 4\beta_3$
$\beta_2 \rightarrow$	β_2	β_4	$-\beta_4$	$-\beta_2$	
$\chi_{(IR)}\,E_g$	2	−2	2	−2	
$\chi_{(IR)}Rv_1$	$2\beta_2$	$-2\beta_4$	$-2\beta_4$	$2\beta_2$	$= 4\beta_2 - 4\beta_4$

From which:

$A_{2u} = 1/2(\beta_1 + \beta_2 + \beta_3 + \beta_4)$

$B_{2u} = 1/2(\beta_1 - \beta_2 + \beta_3 - \beta_4)$

$E_g = 1/\sqrt{2}(\beta_1 - \beta_3),\ 1/\sqrt{2}(\beta_2 - \beta_4)$

APPENDIX 4
Answers to Problems

Chapter 1

1. NH_3: C_3 (passing through N and the mid-point of the H_3 triangle), three σ_v (each containing one N-H bond).

   AsH_5: C_3 and S_3 coincident (along the axial H-As-H unit), three C_2 (along each equatorial As-H bond), σ_h (containing the equatorial AsH_3 unit), three σ_v (each containing the axial H-As-H unit and one equatorial As-H bond).

   cyclo-$B_3N_3H_6$: C_3, S_3 coincident (perpendicular to the molecular plane and passing through the mid-point of the ring), three C_2 (containing pairs of B, N atoms on opposite sides of the ring), σ_h (molecular plane), three σ_v (perpendicular to the molecular plane and each containing a pair of B,N atoms).

   B(H)(F)(Br): the molecule is trigonal planar and only has a mirror plane of symmetry (containing all four atoms). Since there is no main axis the mirror plane is given the symbol σ (without qualification) as the subscripts *h* and *v* have no significance.

   $C_4H_4F_2Cl_2$: *i* (at mid-point of the four-carbon ring), three S_2 (e.g. passing through *i* and perpendicular to the four-carbon plane, plus two others orthogonal to this). $i \equiv$ any one of three orthogonal S_2 axes since $(x, y, z)\rightarrow(-x, -y, -z)$ can be brought about in any one of three ways.

   SiH_4:

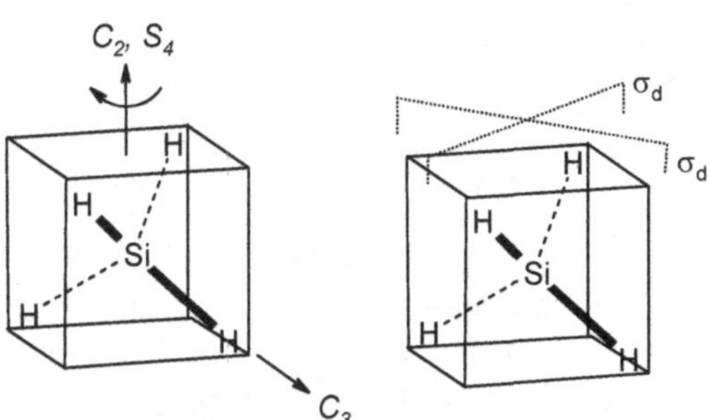

   There are four C_3 axes (lying along each Si-H bond), three C_2 axes (one through the centre of each face of the cube), three S_4 coincident with C_2 and six σ_d (the two shown are replicated across each face of the cube).

2. The pairs of molecules have been chosen to highlight the importance of getting the correct molecular shape as a pre-requisite for assigning symmetry.

   CO_2 is linear, $D_{\infty h}$ while SO_2 is angular and C_{2v}.

Ferrocene (staggered) D_{5d} and ruthenocene (eclipsed) D_{5h}. Both have a C_5 axis and five C_2 axes, but only ruthenocene has a σ_h; ferrocene does have five σ_d.
cis- and *trans*-$Mo(CO)_4Cl_2$ are both octahedral in shape, and have C_{2v} and D_{4h} symmetry, respectively.
$[IF_6]^+$ is perfectly octahedral O_h, while $[IF_6]^-$ is pentagonal pyramidal by VSEPR (like IF_7 but a lone pair in an axial site) and is C_{5v}.
SnCl(F) is angular and C_s, while XeCl(F) is linear and $C_{\infty v}$.
mer- and *fac*- WCl_3F_3 are both octahedral in shape, of C_{2v}, C_{3v} symmetry, respectively.

4. The *cis*-isomer of $CrCl_2(acac)_2$ belongs to the C_2 point group and is chiral and polar, while the *trans*-isomer belongs to the D_{2h} point groups and is neither chiral nor polar.
cyclo-$(Cl_2PN)_4$ belongs to the S_4 point group and is therefore not chiral but does have a dipole.

Chapter 2

1. $\mathbf{A}: \chi = 1 + 1 + 2 = 4 \quad \mathbf{B}: \chi = 2 + (-2) + 3 = 3$

$$\mathbf{C} = \begin{bmatrix} 1 & 2 & 1 \\ 3 & 1 & 1 \\ 1 & 0 & 2 \end{bmatrix} \begin{bmatrix} 2 & 0 & 2 \\ 1 & -2 & 1 \\ 0 & 3 & 3 \end{bmatrix} = \begin{bmatrix} 4 & -1 & 7 \\ 7 & 1 & 10 \\ 2 & 6 & 8 \end{bmatrix} \quad \chi = 4 + 1 + 8 = 13$$

2.

$$\begin{bmatrix} 1 & 0 & 0 & 0 & 0 \\ 0 & -1 & 0 & 0 & 0 \\ 0 & 0 & -1 & 0 & 0 \\ 0 & 0 & 0 & 0 & 1 \\ 0 & 0 & 0 & 1 & 0 \end{bmatrix} \begin{bmatrix} d_{z^2} \\ d_{x^2-y^2} \\ d_{xy} \\ d_{yz} \\ d_{xz} \end{bmatrix} = \begin{bmatrix} d_{z^2} \\ -d_{x^2-y^2} \\ -d_{xy} \\ d_{xz} \\ d_{yz} \end{bmatrix}$$

d_{z^2} transforms onto itself as it lies along the *z* axis.
$d_{x^2-y^2}$ and d_{xy} both turn through 90°, so they remain in their same positions but in each case the + lobes are transformed onto – lobes and *vice versa*. The orbitals thus transform onto the reverse of themselves.
d_{xz} turns through 90° and maps onto d_{yz} while d_{yz} similarly maps onto d_{xz}.

Chapter 3

1.

C_{2v}	E	C_2	$\sigma(xz)$	$\sigma(yz)$
Γ	6	4	–2	0

$= 2A_1 + 3A_2 + B_2$

$A_1 = 1/4[(1 \times 6 \times 1) + (1 \times 4 \times 1) + (1 \times -2 \times 1) + (1 \times 0 \times 1)] = 2$
$A_2 = 1/4[(1 \times 6 \times 1) + (1 \times 4 \times 1) + (1 \times -2 \times -1) + (1 \times 0 \times -1)] = 3$
$B_1 = 1/4[(1 \times 6 \times 1) + (1 \times 4 \times -1) + (1 \times -2 \times 1) + (1 \times 0 \times -1)] = 0$
$B_2 = 1/4[(1 \times 6 \times 1) + (1 \times 4 \times -1) + (1 \times -2 \times -1) + (1 \times 0 \times 1)] = 1$

T_d	E	$8C_3$	$3C_2$	$6S_4$	$6\sigma_d$
Γ	9	3	1	3	3

$= 3A_1 + T_1 + T_2$

$A_1 = 1/24[(1 \times 9 \times 1) + (8 \times 3 \times 1) + (3 \times 1 \times 1) + (6 \times 3 \times 1) + (6 \times 3 \times 1)] = 3$
$A_2 = 1/24[(1 \times 9 \times 1) + (8 \times 3 \times 1) + (3 \times 1 \times 1) + (6 \times 3 \times -1) + (6 \times 3 \times -1)] = 0$
$E = 1/24[(1 \times 9 \times 2) + (8 \times 3 \times -1) + (3 \times 1 \times 2) + (6 \times 3 \times 0) + (6 \times 3 \times 0)] = 0$
$T_1 = 1/24[(1 \times 9 \times 3) + (8 \times 3 \times 0) + (3 \times 1 \times -1) + (6 \times 3 \times 1) + (6 \times 3 \times -1)] = 1$
$T_2 = 1/24[(1 \times 9 \times 3) + (8 \times 3 \times 0) + (3 \times 1 \times -1) + (6 \times 3 \times -1) + (6 \times 3 \times 1)] = 1$

C_{3v}	E	$2C_3$	$3\sigma_v$
Γ	15	0	3

$= 4A_1 + A_2 + 5E$

$A_1 = 1/6[(1 \times 15 \times 1) + (2 \times 0 \times 1) + (3 \times 3 \times 1)] = 4$
$A_2 = 1/6[(1 \times 15 \times 1) + (2 \times 0 \times 1) + (3 \times 3 \times -1)] = 1$
$E = 1/6[(1 \times 15 \times 2) + (2 \times 0 \times -1) + (3 \times 3 \times 0)] = 5$

2.

NH_3 (belongs to the C_{3v} point group)

C_{3v}	E	$2C_3$	$3\sigma_v$
unshifted atoms	4	1	2
$\times \chi_{u.a.}$	3	0	1
Γ_{3N}	12	0	2

Using the reduction formula:

$\Gamma_{3N} = 3A + A_2 + 4E$ (= 3N i.e. 12, which is correct remembering E means doubly degenerate)

$\Gamma_{trans+rot} = A_1 + A_2 + 2E$ (= 6)

$\Gamma_{vib} = 2A_1 + 2E$ (= 3N – 6 i.e. 6, again remembering E means doubly degenerate)

cis-N_2H_2 (belongs to the C_{2v} point group; let *yz* be the molecular plane)

C_{2v}	E	C_2	$\sigma(xz)$	$\sigma(yz)$
unshifted atoms	4	0	0	4
$\times \chi_{u.a.}$	3	–1	1	1
Γ_{3N}	12	0	0	4

Which reduces to:

$\Gamma_{3N} = 4A_1 + 2A_2 + 2B_1 + 4B_2$ (= 3N i.e. 12)
$\Gamma_{trans+rot} = A_1 + A_2 + 2B_1 + 2B_2$ (= 6)
$\Gamma_{vib} = 3A_1 + A_2 + 2B_2$ (= 3N – 6 i.e. 6)

SO_3 (belongs to the D_{3h} point group)

D_{3h}	E	$2C_3$	$3C_2$	σ_h	$2S_3$	$3\sigma_v$
unshifted atoms	4	1	2	4	1	2
$\times \chi_{u.a.}$	3	0	–1	1	–2	1
Γ_{3N}	12	0	–2	4	–2	2

Which reduces to:

$\Gamma_{3N} = A_1' + A_2' + 3E' + 2A_2'' + E''$ (= 3N i.e. 12)
$\Gamma_{trans+rot} = A_2' + E' + A_2'' + E''$ (= 6)
$\Gamma_{vib} = A_1' + 2E' + A_2''$ (= 3N – 6 i.e. 6)

3.

C_{2h}	E	C_2	i	σ_h
unshifted atoms	10	0	0	10
$\times \chi_{u.a.}$	3	–1	–3	1
Γ_{3N}	30	0	0	10

Which reduces to:

$\Gamma_{3N} = 10A_g + 5B_g + 5A_u + 10B_u$ (= 3N i.e. 30)
$\Gamma_{trans+rot} = A_g + 2B_g + A_u + 2B_u$ (= 6)
$\Gamma_{vib} = 9A_g + 3B_g + 4A_u + 8B_u$ (= 3N – 6 i.e. 24)

Chapter 4

1. $[ClO_4]^-$ belongs to the T_d point group.
   A_1 : same symmetry as $x^2 + y^2 + z^2$, thus Raman-only active; as this representation is perfectly symmetrical the Raman band will be polarised.
   E : has the symmetry of $2z^2 - x^2 - y^2$, $x^2 - y^2$ and is also Raman-only active; this is an unsymmetrical modes and is Raman depolarised.
   T_2 : this has the symmetry of $\mathbf{T_x}$, $\mathbf{T_y}$, $\mathbf{T_z}$ and xy, yz, zx and is thus both infrared and Raman active; the Raman band is depolarised as it is an unsymmetrical mode.

2. $[BrF_2]^-$: VSEPR predicts this is linear and thus belongs to the $D_{\infty h}$ point group. The anion has an inversion centre on the bromine and thus its vibrational spectrum should have no coincidences between the infrared and Raman spectra. It is spectrum **A**.

$[BrF_2]^+$: VSEPR predicts an angular structure i.e. C_{2v} point group. This has no inversion centre and thus coincidences between the infrared and Raman spectra are to be expected. This is spectrum **B**.

Chapter 5

2.

C_{3v}	E	$2C_3$	$3\sigma_v$	
Γ_{3N}	15	0	3	$= 4A_1 + A_2 + 5E$

$\Gamma_{trans+rot} = A_1 + A_2 + 2E$

$\Gamma_{vib} = 3A_1 + 3E$

Both A_1 and E are infrared and Raman active; A_1 is Raman pol, E is Raman depol.

$\Gamma_{P\text{-}O}$	1	1	1	$= A_1$	
$\Gamma_{P\text{-}Cl}$	3	0	1	$= A_1 + E$	
$\Gamma_{Cl\text{-}P\text{-}Cl/Cl\text{-}P\text{-}O}$	6	0	2	$= 2A_1 + 2E$	One A_1 is redundant.

Infrared (liquid; cm^{-1})	Raman (liquid ; cm^{-1})		
1292	1290 (pol)	A_1	P=O stretch
580	581 (depol)	E	P-Cl stretch
487	486 (pol)	A_1	P-Cl stretch
340	337 (depol)	E	deformation
267	267 (pol)	A_1	deformation
not accessible	193 (depol)	E	deformation

3.

C_{2h}	E	C_2	i	σ_v	
Γ_{3N}	12	0	0	4	$= 4A_g + 2B_g + 2A_u + 4B_{2u}$

$\Gamma_{trans+rot} = A_g + 2B_g + A_u + 2B_u$

$\Gamma_{vib} = 3A_g + A_u + 2B_u$

A_g is Raman-only active and pol; A_u, B_u both infrared-only active.

$\Gamma_{N\text{-}N}$	1	1	1	1	$= A_g$
$\Gamma_{N\text{-}F}$	2	0	0	2	$= A_g + B_u$
$\Gamma_{F\text{-}N\text{-}N}$	2	0	0	2	$= A_g + B_u$
$\Gamma_{out\text{-}of\text{-}plane}$	2	0	0	–2	$= A_u + B_g$

B_g is redundant (corresponds to $\mathbf{R_x}$, $\mathbf{R_y}$)

Infrared (gas, cm^{-1})	Raman (gas, cm^{-1})		
	1636 (pol)	A_g	N=N stretch
	1010 (pol)	A_g	N-F stretch
989		B_u	N-F stretch
	592 (pol)	A_g	F-N-N deformation
412		A_u *or* B_u	out-of-plane deformation
360		A_u *or* B_u	out-of-plane deformation

Decreasing energy : N=N stretch > N-F stretch > bending modes

Chapter 6

1.

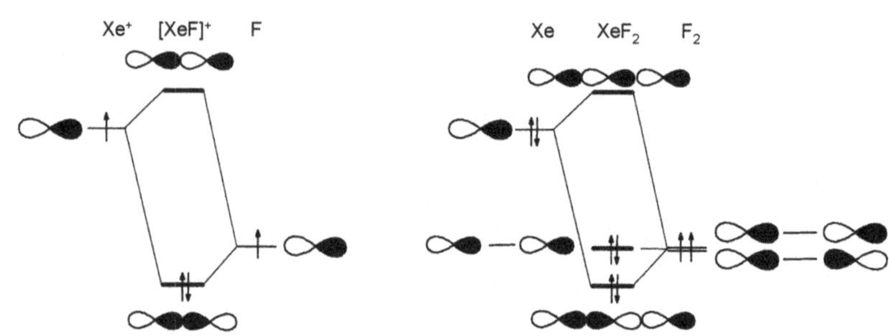

In $[XeF]^+$, Xe has one electron in p_z (allowing for the loss of an electron to generate the cation), while fluorine also has one. The Xe-F bond order is 1.

In XeF_2, the total of four electrons (Xe: 2; F: 2 × 1) are distributed equally over the bonding and non-bonding MOs. There is a total bond order of 1 i.e. 0.5 per Xe-F bond.

3.

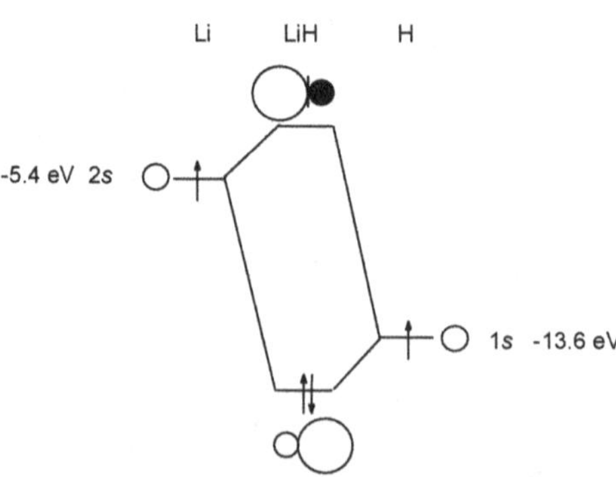

The two electrons in the bonding MO are localised on hydrogen, consistent with a bond polarity of $Li^{\delta+}$-$H^{\delta-}$.

4.

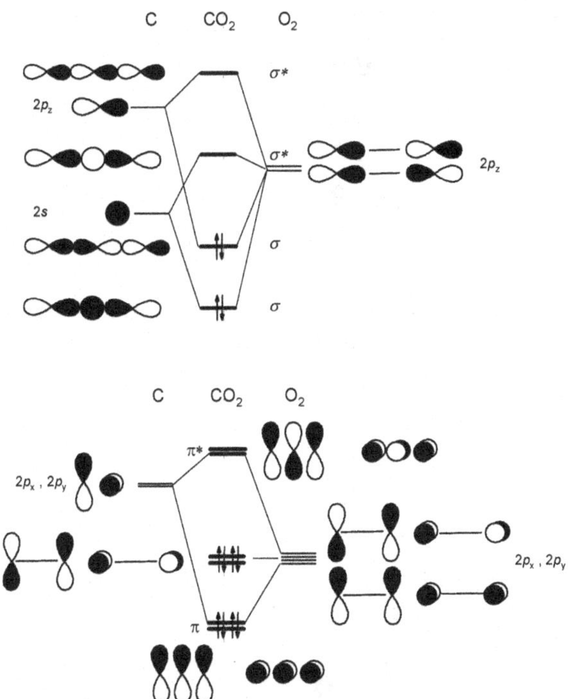

For the σ-bonds, one of the two combinations for the p_z orbitals on oxygen combines with the *s*-orbital on carbon, while the other combines with the carbon p_z; both bonding and anti-bonding combinations are formed.

For the π-bonds, the in-phase combinations formed by both p_x and p_y on oxygen combine with their equivalent orbital on carbon to give a pair of π-bonding and a pair of π* anti-bonding MOs. The out-of-phase pair of oxygen combinations have no match with an AO on carbon and are non-bonding.

There are a total of twelve bonding electrons (C: $2s^2\,2p^2$; O $2p^4$) which fill the two σ bonding MOs, the two π-bonding MOs and the two non-bonding MOs. The total bond order is 4 i.e. a bond order of 2 per C-O unit.

Chapter 7

1.

C_{2v}	E	C_2	$\sigma(xz)$	$\sigma(yz)$	
$\Gamma_{O\,px}$	2	0	0	–2	$= a_2 + b_1$
$\Gamma_{N\,2px}$	1	–1	1	–1	$= b_1$

The symmetry of the nitrogen p_x could have been read directly from the character table : same symmetry as $\mathbf{T_x}$.

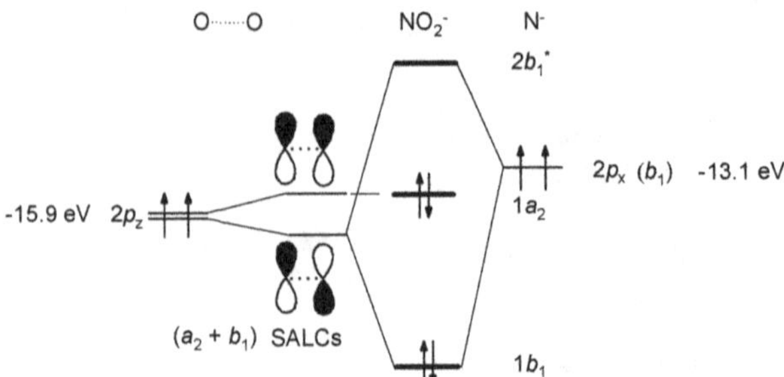

Each atom contributes one π-electron, with an additional π-electron for the negative charge; this has been arbitrarily assigned to nitrogen in the MO diagram.

The π-bond order is 1 i.e. 0.5 per N-O π-bond.

Chapter 8

1. BF_3 can be considered as sp^2 hybridised at boron, leaving an empty p_z orbital free for π-bonding with the filled p_z lone pair orbitals on fluorine.

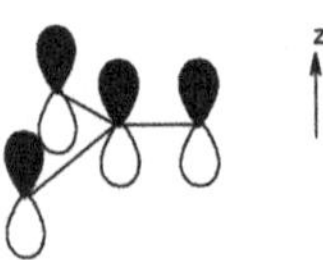

D_{3h}	E	$2C_3$	$3C_2$	σ_h	$2S_3$	$3\sigma_v$	
$\Gamma_{F\,pz}$	3	0	−1	−3	0	1	$a_2'' + e''$

The p_z AO on boron has a_2'' symmetry (symmetry of $\mathbf{T_z}$).

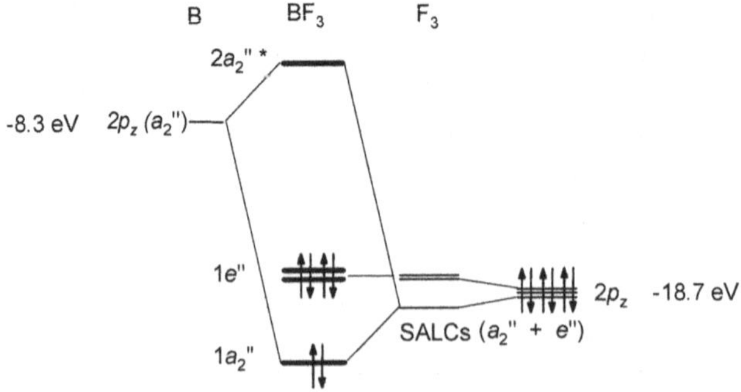

The p_z on boron is vacant while each p_z on fluorine contains two electrons.

π-Bond order = 1 i.e. 0.333 per B-F bond.

Boron in BF_3 should be a strong Lewis acid as (*i*) it has a vacant *p*-orbital capable of accepting an electron pair and (*ii*) it should be strongly $B^{\delta+}$ by virtue of the three very electronegative fluorines. However, intramolecular π-bond formation means that the p_z on boron is not vacant and is thus only a weak Lewis acid.

6.

D_{6h}	E	$2C_6$	$2C_3$	C_2	$3C_2'$	$3C_2''$	i	$2S_3$	$2S_6$	σ_h	$3\sigma_d$	$3\sigma_v$
$\Gamma_{C\,pz}$	6	0	0	0	–2	0	0	0	0	–6	0	2

$= b_{2g} + e_{1g} + a_{2u} + e_{2u}$ (this is reasonable as we expect six MOs)

The e_{1g} and e_{2u} MOs each comprise a pair of SALCs with either one node or two nodes, which can be distinguished by the subscripts *g* and *u*. Of the two singly-degenerate MOs, a_{2u} is the one with no nodes (*u*) while the most anti-bonding MO, with three nodes, is b_{2g} (*g*).

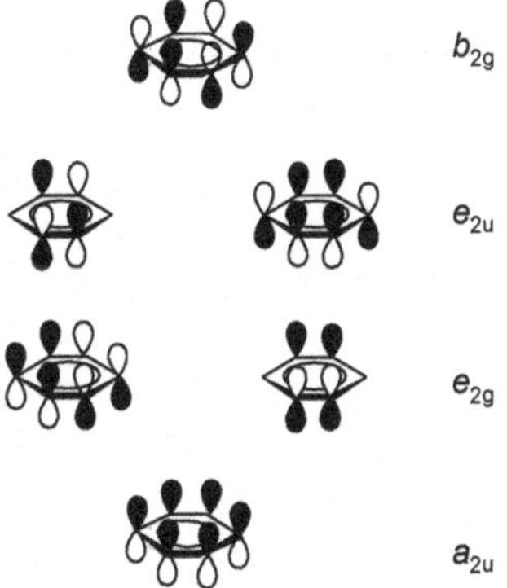

Symmetry labels for the a_{2u} and b_{2g} SALCs can be checked by treating each of them as a complete unit and counting 1, 0, -1 if they are unmoved, moved or reversed under each of the D_{6h} symmetry operations; the pairs of SALCs described by the *e* labels cannot be analysed in this way.

Chapter 9

1.

C_{6v}	E	$2C_6$	$2C_3$	C_2	$3\sigma_v$	$3\sigma_d$
$\Gamma_{6\ C\,pz}$	6	0	0	0	2	0

$= a_1 + b_1 + e_1 + e_2$

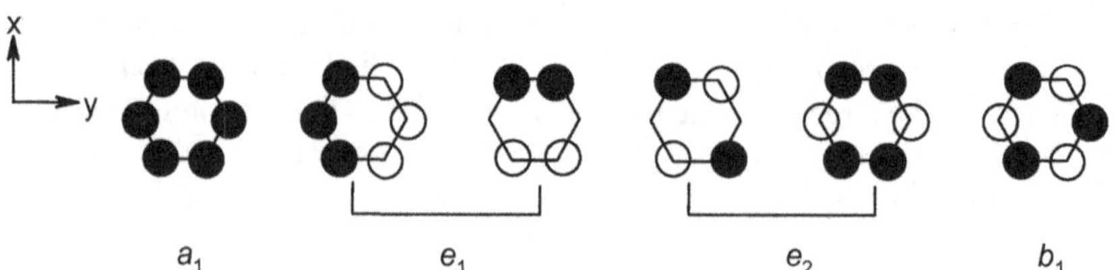

The e_1 and e_2 pairs can be distinguished by their matches with the *d*-orbitals on iron.

AOs on iron:

d_{z^2} : a_1
$d_{x^2-y^2}$: e_2
d_{xy} : e_2
d_{xz} : e_1
d_{yz} : e_1

5.

D_{4h}	E	$2C_4$	C_2	$2C_2'$	$2C_2''$	i	$2S_4$	σ_h	$2\sigma_v$	$2\sigma_d$
Γ_{Fp}	4	0	0	2	0	0	0	4	2	0

This reduces to: $a_{1g} + b_{1g} + e_u$ (correct, as we expect four SALCs)

The symmetry labels for the valence orbitals on xenon, read from the character table are:

s : a_{1g} d_{xz}, d_{yz} : e_g
p_x, p_y : e_u $d_{x^2-y^2}$: b_{1g}
p_z : a_{2u} d_{z^2} : a_{1g}
d_{xy} : b_{2g}

MO combinations:

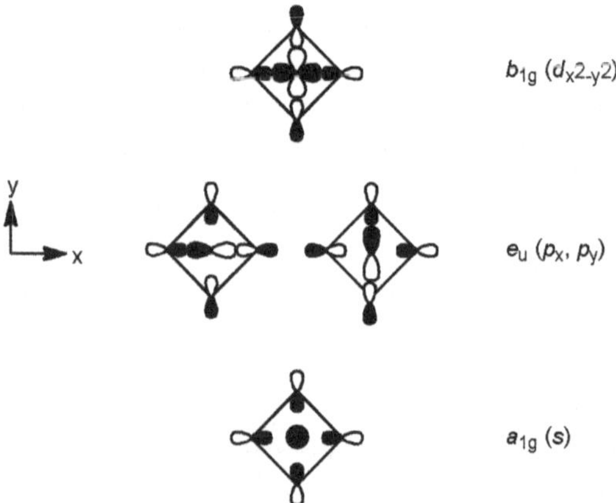

Note: In the e_u SALCs two *p* orbitals in each case make non-bonding contributions and can be ignored.

Chapter 10

2.

C_{2h}	E	C_2	i	σ_h
$\Gamma_{6\,C\,pz}$	6	0	0	2

$= 2a_g + b_g + a_u + 2b_u$

The three SALCs associated with each allyl group are:

Combining identical pairs, both in-phase and out-of-phase, gives the following, exemplified by SALCs associated with combinations of the in-phase set of three p_z AOs:

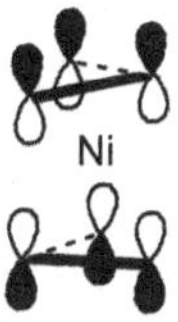

ψ_1

ψ_2

ψ_3 - ψ_6 are shown in the Table, below.

C_{2h}	E	C_2	i	σ_h	
ψ_1	1	1	1	1	$= a_g$
ψ_2	1	−1	−1	1	$= b_u$
ψ_3	1	1	−1	−1	$= a_u$
ψ_4	1	−1	1	−1	$= b_g$
ψ_5	1	−1	−1	1	$= b_u$
ψ_6	1	1	1	1	$= a_g$

The AO column (below) shows all symmetry-allowed matches between nickel AOs and ligand SALCs; the best matches are shown without parentheses. This analysis is a simplification in that further mixing, e.g. of all AOs / SALCs of a_g or b_u symmetry, will take place to some extent.

		SALC	AO	Label
	ψ_1	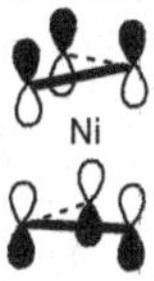	s, $d_{x2\text{-}y2}$ (d_{z2}, d_{xy})	a_g
	ψ_2		p_x (p_y)	b_u

Question 2, continued.

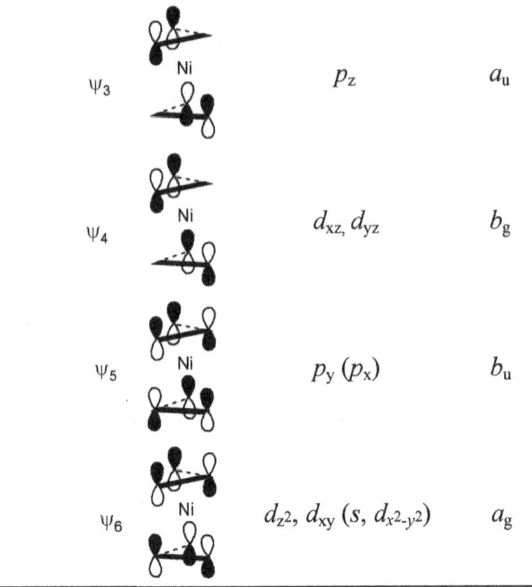

Chapter 11

1. The ground state $(a_1)^2(e)^4(a_1)^2$ has $^1A_1$ symmetry. The excited states and their symmetries are:

$(a_1)^2(e)^4(a_1)^1(a_1*)^1 = A_1 \times A_1 \times A_1 \times A_1 = {}^1A_1$ or ${}^3A_1$
$(a_1)^2(e)^4(a_1)^1(e*)^1 = A_1 \times A_1 \times A_1 \times E = {}^1E$ or ${}^3E$

Only singlet excited states will afford spin-allowed transitions. The symmetry of μ is $A_1 + E$, so the transition integrals are:

$(a_1)^2(e)^4(a_1)^2 \rightarrow (a_1)^2(e)^4(a_1)^1(a_1*)^1$: $A_1 \times A_1 \times A_1$ (=A_1) or $A_1 \times E \times A_1$
$(a_1)^2(e)^4(a_1)^2 \rightarrow (a_1)^2(e)^4(a_1)^1(e*)^1$: $A_1 \times A_1 \times E$ or $A_1 \times E \times E$ (contains A_1)

So, both $^1A_1 \rightarrow {}^1A_1$ and $^1A_1 \rightarrow {}^1E$ transitions are symmetry allowed.

3. The ground state $(a_u)^2(b_g)^2$ has ^1A_g symmetry. The excited states and their symmetries are:

$(a_u)^2(b_g)^1(a_u*)^1 = A_g \times B_g \times A_u = {}^1B_u$ or ${}^3B_u$
$(a_u)^2(b_g)^1(b_g*)^1 = A_g \times B_g \times B_g = {}^1A_g$ or ${}^3A_g$

Only singlet excited states will afford spin-allowed transitions. The symmetry of μ is $A_u + B_u$, so the transition integrals are:

$(a_u)^2(b_g)^2 \rightarrow (a_u)^2(b_g)^1(a_u*)^1 = A_g \times A_u \times B_u\ (= B_g)$ or $A_g \times B_u \times B_u\ (= A_g)$
$(a_u)^2(b_g)^2 \rightarrow (a_u)^2(b_g)^1(b_g*)^1 = A_g \times A_u \times A_g\ (= A_u)$ or $A_g \times B_u \times A_g\ (= B_u)$

Thus, only the ${}^1A_g \rightarrow {}^1B_u\ (b_g \rightarrow a_u*)$ transition is symmetry-allowed as the transition integral contains A_g.

4. The ground state $(a_{2u})^2(e_g)^4$ has ${}^1A_{1g}$ symmetry. The excited states and their symmetries are:

$(a_{2u})^2(e_g)^3(b_{2u}*)^1 = A_{1g} \times E_g \times B_{2u} = E_u$; $(e_g)^3 \equiv (e_g)^1$ by the hole formalism
$(a_{2u})^1(e_g)^4(b_{2u}*)^1 = A_{2u} \times A_{1g} \times B_{2u} = B_{1g}$

Only singlet excited states will afford spin-allowed transitions. The symmetry of μ is $A_{2u} + E_u$, so the transition integrals are:

$(a_{2u})^2(e_g)^4 \rightarrow (a_{2u})^2(e_g)^3(b_{2u}*)^1 = A_{1g} \times A_{2u} \times E_u\ (= E_g)$
or $= A_{1g} \times E_u \times E_u$ (contains A_{1g})
$(a_{2u})^2(e_g)^4 \rightarrow (a_{2u})^1(e_g)^4(b_{2u}*)^1 = A_{1g} \times A_{2u} \times B_{1g}\ (= B_{2u})$ or $A_{1g} \times E_u \times B_{1g}\ (= E_u)$

Thus, only the ${}^1A_{1g} \rightarrow {}^1E_u\ [(a_{2u})^2(e_g)^4 \rightarrow (a_{2u})^2(e_g)^3(b_{2u}*)^1]$ transition is symmetry-allowed as the transition integral contains A_{1g}.

Chapter 12

1. The degeneracy of the configuration is given by the product of the degeneracies of its components:

$(s)^1 = 2$ and $(p)^1 = 6$ (using eqn 12.1), so the total degeneracy for $(s)^1(p)^1 = 12$

For the $(s)^1$, $m_l = 0$ and $m_s = ½$ and for $(p)^1$ $m_l = 1, 0$ or -1 and $m_s = ½$

Maximum $L = 1$ ($m_l = 0$ for s and 1 for p) and maximum $S = ½ + ½ = 1$.

The grid for the microstates is as follows:

M_L	M_S 1	M_S 0	M_S -1
1	**$(1^+,0^+)$**	**$(1^+,0^-)$** $(1^-,0^+)$	**$(1^-,0^-)$**
0	**$(0^+,0^+)$**	**$(0^+,0^-)$** $(0^-,0^+)$	**$(0^-,0^-)$**
-1	**$(-1^+,0^+)$**	**$(-1^+,0^-)$** $(-1^-,0^+)$	**$(-1^-,0^-)$**

The ground term corresponds to maximum L, S and is ${}^3P$ (M_L= 1, 0, -1; M_S = 1, 0, -1); this corresponds to the nine microstates in bold. Of the microstates remaining, the one with maximum L, S is $(1^{-},0^{+})$ which arises from a ^1P term and accounts for the remaining three microstates.

4. From Table 12.6 (Section 12.5):

$(t_{2g})^4(e_g)^1 = {}^3T_{1g} \times {}^2E_g = T_{1g} + T_{2g}$ ignoring spin multiplicities

For $^3T_{1g}$ $S = 1$ $(2S + 1 = 3)$ and for ^2E_g $S = ½$, so maximum $S = {}^3/_2$ and $2S + 1 = 4$, thus, $(t_{2g})^4(e_g)^1 = {}^3T_{1g} \times {}^2E_g = {}^4T_{1g} + {}^4T_{2g}$

Similarly, $(t_{2g})^3(e_g)^2 = {}^4A_{2g} \times {}^3A_{2g} = {}^6A_{1g}$

5. The ground term has maximum S (= ½ + ½) and maximum L. $L = 6$ (3 + 3) is incompatible with maximum S, so the maximum L is 5 (3 + 2). The ground term is ^3H. Using Eqn. 12.2 – 12.6 with "+" in the formulae (f-electrons are u, so $f^2 = u \times u = g$) gives:

O_h	E	$8C_3$	$6C_2$	$6C_4$	$3C_2^{\,a}$	i	$6S_4$	$8S_6$	$3\sigma_h$	$6\sigma_d$
Γ_S	11	-1	-1	1	-1	11	1	-1	-1	-1

$= E_g + 2T_{1g} + T_{2g}$

^3H splits into $^3E_g + 2\,{}^3T_{1g} + {}^3T_{2g}$; 33 microstates, as given by $(2S + 1)(2L + 1)$.

Chapter 13

1. The symmetry of the dipole moment is T_{1u} (O_h) or T_2 (T_d), so the transition integrals are:

$$O_h : {}^2E_g \leftarrow {}^2T_{2g} : \; T_{2g} \times T_{1u} \times E_g = (A_{2u} + E_u + T_{1u} + T_{2u}) \times E_g$$
$$= E_u + A_{1u} + A_{2u} + E_u + T_{1u} + T_{2u} + T_{1u} + T_{2u}$$

$$T_d : {}^2T_2 \leftarrow {}^2E : E \times T_2 \times T_2 = (T_1 + T_2) \times T_2 = A_2 + E + T_1 + T_2 + A_1 + E + T_1 + T_2$$

Only $^2T_2 \leftarrow {}^2E$ spans the totally symmetry irreducible representation for its point group (A_1) so is allowed while $^2E_g \leftarrow {}^2T_{2g}$, which does not span A_{1g}, is forbidden.

3. $(t_{2g})^6$ has $^1A_{1g}$ symmetry as all the t_{2g} orbitals are filled.
$(t_{2g})^5(e_g)^1 \equiv (t_{2g})^1(e_g)^1$ by the hole formalism $= T_{2g} \times E_g = T_{1g} + T_{2g}$

For $(t_{2g})^5(e_g)^1$ both singlet and triplet states are possible but only the singlet excited states will afford spin-allowed transitions, which therefore are:

$$^1T_{1g} \leftarrow {}^1A_{1g} \quad \text{and} \quad {}^1T_{2g} \leftarrow {}^1A_{1g}$$

4. $(e_g)^1 = {}^2E_g$, while from Table 12.6 in Section 12.5:

$$(t_{2g})^3 = {}^4A_{2g} + {}^2E_g + {}^2T_{1g} + {}^2T_{2g}$$

$$(t_{2g})^3(e_g)^1 = ({}^4A_{2g} + {}^2E_g + {}^2T_{1g} + {}^2T_{2g}) \times {}^2E_g$$

Considering each binary direct product in turn:

${}^4A_{2g} \times {}^2E_g = {}^5E_g + {}^3E_g$ (see Table 13.2 for combinations of spin multiplicities)

${}^2E_g \times {}^2E_g = {}^3A_{1g} + {}^3A_{2g} + {}^3E_g + {}^1A_{1g} + {}^1A_{2g} + {}^1E_g$

${}^2T_{1g} \times {}^2E_g = {}^3T_{1g} + {}^3T_{2g} + {}^1T_{1g} + {}^1T_{2g}$; ${}^2T_{2g} \times {}^2E_g$ gives the same result.

Overall:
$(t_{2g})^3(e_g)^1 = {}^5E_g + 2{}^3E_g + E_g + {}^3A_{1g} + {}^3A_{2g} + {}^1A_{1g} + {}^1A_{2g} + 2{}^3T_{1g} + 2{}^3T_{2g} + 2{}^1T_{1g} + 2{}^1T_{2g}$

Degeneracy: $(t_{2g})^3 = 6!/(3! \times 3!) = 20$; $(e_g)^1 = 4$ so $(t_{2g})^3(e_g)^1 = 80$

$(t_{2g})^3(e_g)^1 = 10 + 12 + 2 + 3 + 3 + 1 + 1 + 18 + 18 + 6 + 6 = 80$

Spin allowed transitions from ${}^3T_{1g}$ ground state are:

${}^3E_g \leftarrow {}^3T_{1g}$ ${}^3A_{1g} \leftarrow {}^3T_{1g}$ ${}^3A_{2g} \leftarrow {}^3T_{1g}$
${}^3T_{1g} \leftarrow {}^3T_{1g}$ ${}^3T_{2g} \leftarrow {}^3T_{1g}$

APPENDIX 5
Selected Character Tables

C_s	E	σ_h		
A′	1	1	$\mathbf{T_x, T_y, R_z}$	x^2, y^2, z^2, xy
A′′	1	-1	$\mathbf{T_z, R_x, R_y}$	yz, xz

C_{2v}	E	C_2	$\sigma(xz)$	$\sigma(yz)$		
A_1	1	1	1	1	$\mathbf{T_z}$	x^2, y^2, z^2
A_2	1	1	-1	-1	$\mathbf{R_z}$	xy
B_1	1	-1	1	-1	$\mathbf{T_x, R_y}$	xz
B_2	1	-1	-1	1	$\mathbf{T_y, R_x}$	yz

C_{3v}	E	$2C_3$	$3\sigma_v$		
A_1	1	1	1	$\mathbf{T_z}$	$x^2 + y^2, z^2$
A_2	1	1	-1	$\mathbf{R_z}$	
E	2	-1	0	$(\mathbf{T_x, T_y}), (\mathbf{R_x, R_y})$	$(x^2 - y^2, xy), (yz, xz)$

C_{4v}	E	$2C_4$	C_2	$2\sigma_v$	$2\sigma_d$		
A_1	1	1	1	1	1	$\mathbf{T_z}$	$x^2 + y^2, z^2$
A_2	1	1	1	-1	-1	$\mathbf{R_z}$	
B_1	1	-1	1	1	-1		$x^2 - y^2$
B_2	1	-1	1	-1	1		xy
E	2	0	-2	0	0	$(\mathbf{T_x, T_y}), (\mathbf{R_x, R_y})$	(yz, xz)

C_{6v}	E	$2C_6$	$2C_3$	C_2	$3\sigma_v$	$3\sigma_d$		
A_1	1	1	1	1	1	1	$\mathbf{T_z}$	x^2+y^2,z^2
A_2	1	1	1	1	-1	-1	$\mathbf{R_z}$	
B_1	1	-1	1	-1	1	-1		
B_2	1	-1	1	-1	-1	1		
E_1	2	1	-1	-2	0	0	$(\mathbf{T_x,T_y}), (\mathbf{R_x,R_y})$	(xz, yz)
E_2	2	-1	-1	2	0	0		$(x^2 - y^2, xy)$

C_{2h}	E	C_2	i	σ_h		
A_g	1	1	1	1	$\mathbf{R_z}$	x^2, y^2, z^2, xy
B_g	1	-1	1	-1	$\mathbf{R_x}, \mathbf{R_y}$	yz, xz
A_u	1	1	-1	-1	$\mathbf{T_z}$	
B_u	1	-1	-1	1	$\mathbf{T_x}, \mathbf{T_y}$	

D_{2h}	E	$C_2(z)$	$C_2(y)$	$C_2(x)$	i	$\sigma(xy)$	$\sigma(xz)$	$\sigma(yz)$		
A_g	1	1	1	1	1	1	1	1		x^2, y^2, z^2
B_{1g}	1	1	-1	-1	1	1	-1	-1	$\mathbf{R_z}$	xy
B_{2g}	1	-1	1	-1	1	-1	1	-1	$\mathbf{R_y}$	xz
B_{3g}	1	-1	-1	1	1	-1	-1	1	$\mathbf{R_x}$	yz
A_u	1	1	1	1	-1	-1	-1	-1		
B_{1u}	1	1	-1	-1	-1	-1	1	1	$\mathbf{T_z}$	
B_{2u}	1	-1	1	-1	-1	1	-1	1	$\mathbf{T_y}$	
B_{3u}	1	-1	-1	1	-1	1	1	-1	$\mathbf{T_x}$	

D_{3h}	E	$2C_3$	$3C_2$	σ_h	$2S_3$	$3\sigma_v$		
A_1'	1	1	1	1	1	1		$x^2 + y^2, z^2$
A_2'	1	1	-1	1	1	-1	$\mathbf{R_z}$	
E'	2	-1	0	2	-1	0	$(\mathbf{T_x}, \mathbf{T_y})$	$(x^2 - y^2, xy)$
A_1''	1	1	1	-1	-1	-1		
A_2''	1	1	-1	-1	-1	1	$\mathbf{T_z}$	
E''	2	-1	0	-2	1	0	$(\mathbf{R_x}, \mathbf{R_y})$	(xz, yz)

D_{4h}	E	$2C_4$	C_2	$2C_2'$	$2C_2''$	i	$2S_4$	σ_h	$2\sigma_v$	$2\sigma_d$		
A_{1g}	1	1	1	1	1	1	1	1	1	1		x^2+y^2, z^2
A_{2g}	1	1	1	-1	-1	1	1	1	-1	-1	$\mathbf{R_z}$	
B_{1g}	1	-1	1	1	-1	1	-1	1	1	-1		$x^2 - y^2$
B_{2g}	1	-1	1	-1	1	1	-1	1	-1	1		xy
E_g	2	0	-2	0	0	2	0	-2	0	0	$(\mathbf{R_x}, \mathbf{R_y})$	(xz, yz)
A_{1u}	1	1	1	1	1	-1	-1	-1	-1	-1	-	
A_{2u}	1	1	1	-1	-1	-1	-1	-1	1	1	$\mathbf{T_z}$	
B_{1u}	1	-1	1	1	-1	-1	1	-1	-1	1		
B_{2u}	1	-1	1	-1	1	-1	1	-1	1	-1		
E_u	2	0	-2	0	0	-2	0	2	0	0	$(\mathbf{T_x}, \mathbf{T_y})$	

D_{5h}	E	$2C_5$	$2C_5^2$	$5C_2$	σ_h	$2S_5$	$2S_5^3$	$5\sigma_v$		
A_1'	1	1	1	1	1	1	1	1		x^2+y^2, z^2
A_2'	1	1	1	-1	1	1	1	-1	R_z	
E_1'	2	2cos72	2cos144	0	2	2cos72	2cos144	0	(T_x, T_y)	
E_2'	2	2cos144	2cos72	0	2	2cos144	2cos72	0		(x^2-y^2, xy)
A_1''	1	1	1	1	-1	-1	-1	-1		
A_2''	1	1	1	-1	-1	-1	-1	1	T_z	
E_1''	2	2cos72	2cos144	0	-2	-2cos72	-2cos144	0	(R_x,R_y)	(xz, yz)
E_2''	2	2cos144	2cos72	0	-2	-2cos144	-2cos72	0		

D_{6h}	E	$2C_6$	$2C_3$	C_2	$3C_2'$	$3C_2''$	i	$2S_3$	$2S_6$	σ_h	$3\sigma_d$	$3\sigma_v$		
A_{1g}	1	1	1	1	1	1	1	1	1	1	1	1		x^2+y^2, z^2
A_{2g}	1	1	1	1	-1	-1	1	1	1	1	-1	-1	R_z	
B_{1g}	1	-1	1	-1	1	-1	1	-1	1	-1	1	-1		
B_{2g}	1	-1	1	-1	-1	1	1	-1	1	-1	-1	1		
E_{1g}	2	1	-1	-2	0	0	2	1	-1	-2	0	0	(R_x,R_y)	(xz, yz)
E_{2g}	2	-1	-1	2	0	0	2	-1	-1	2	0	0		(x^2-y^2, xy)
A_{1u}	1	1	1	1	1	1	-1	-1	-1	-1	-1	-1		
A_{2u}	1	1	1	1	-1	-1	-1	-1	-1	-1	1	1	T_z	
B_{1u}	1	-1	1	-1	1	-1	-1	1	-1	1	-1	1		
B_{2u}	1	-1	1	-1	-1	1	-1	1	-1	1	1	-1		
E_{1u}	2	1	-1	-2	0	0	-2	-1	1	2	0	0	(T_x, T_y)	
E_{2u}	2	-1	-1	2	0	0	-2	1	1	-2	0	0		

D_{5d}	E	$2C_5$	$2C_5^2$	$5C_2$	i	$2S_{10}^3$	$2S_{10}$	$5\sigma_d$		
A_{1g}	1	1	1	1	1	1	1	1		x^2+y^2, z^2
A_{2g}	1	1	1	-1	1	1	1	-1	R_z	
E_{1g}	2	2cos72	2cos144	0	2	2cos72	2cos144	0	(R_x, R_y)	(xz, yz)
E_{2g}	2	2cos144	2cos72	0	2	2cos144	2cos72	0		(x^2-y^2, xy)
A_{1u}	1	1	1	1	-1	-1	-1	-1		
A_{2u}	1	1	1	-1	-1	-1	-1	1	T_z	
E_{1u}	2	2cos72	2cos144	0	-2	-2cos72	-2cos144	0	(T_x, T_y)	
E_{2u}	2	2cos144	2cos72	0	-2	-2cos144	-2cos72	0		

O_h	E	$8C_3$	$6C_2$	$6C_4$	$3C_2$ $(=C_4^2)$	i	$6S_4$	$8S_6$	$3\sigma_h$	$6\sigma_d$		
A_{1g}	1	1	1	1	1	1	1	1	1	1		$x^2+y^2+z^2$
A_{2g}	1	1	-1	-1	1	1	-1	1	1	-1		
E_g	2	-1	0	0	2	2	0	-1	2	0		$(2z^2-x^2-y^2, x^2-y^2)$
T_{1g}	3	0	-1	1	-1	3	1	0	-1	-1	(R_x, R_y, R_z)	
T_{2g}	3	0	1	-1	-1	3	-1	0	-1	1		(xy, xz, yz)
A_{1u}	1	1	1	1	1	-1	-1	-1	-1	-1		
A_{2u}	1	1	-1	-1	1	-1	1	-1	-1	1		
E_u	2	-1	0	0	2	-2	0	1	-2	0		
T_{1u}	3	0	-1	1	-1	-3	-1	0	1	1	(T_x, T_y, T_x)	
T_{2u}	3	0	1	-1	-1	-3	1	0	1	-1		

T_d	E	$8C_3$	$3C_2$	$6S_4$	$6\sigma_d$		
A_1	1	1	1	1	1		$x^2+y^2+z^2$
A_2	1	1	1	-1	-1		
E	2	-1	2	0	0		$(2z^2- x^2-y^2, x^2-y^2)$
T_1	3	0	-1	1	-1	$(\mathbf{R_x}, \mathbf{R_y}, \mathbf{R_z})$	
T_2	3	0	-1	-1	1	$(\mathbf{T_x}, \mathbf{T_y}, \mathbf{T_z})$	(xy, xz, yz)

$D_{\infty h}$	E	$2C_\infty^\Phi$	....	$\infty\sigma_v$	i	$2S_\infty^\Phi$	....	∞C_2		
Σ_g^+	1	1	...	1	1	1	....	1		x^2+y^2, z^2
Σ_g^-	1	1	....	-1	1	1	....	-1	$\mathbf{R_z}$	
Π_g	2	2cosΦ	....	0	2	-2cosΦ	....	0	$(\mathbf{R_x},\mathbf{R_y})$	(xz, yz)
Δ_g	2	2cos2Φ	....	0	2	2cos2Φ	....	0		(x^2-y^2, xy)
-	-	-	-	-	-	-	-	-		
Σ_u^+	1	1	....	1	-1	-1	....	-1	$\mathbf{T_z}$	
Σ_u^-	1	1	....	-1	-1	-1	....	1		
Π_u	2	2cosΦ	....	0	-2	2cosΦ	....	0	$(\mathbf{T_x}, \mathbf{T_y})$	
Δ_u	2	2cos2Φ	....	0	-2	-2cos2Φ	....	0		

INDEX

Ammonia
  MO diagram 91
Asymmetry 14
Atomic orbitals
  as basis set 22, 80
  symmetry, central atom 81, 87, 90, 101, 102, 111
Axis
  alternating 8
  improper 8, 14
  rotation 4, 14
  three-fold 4

Basis set
  definition 20
Beer-Lambert law 152
Bending vectors 40, 57, 63
Bond polarity 77
Borane
  MO diagram 90

Character Tables
  definition 23, 26
  selected tables 218
Charge-transfer bands 156
Chirality 14
CFSE 104, 106
Configuration 133, 148, 150
Crystal field theory 98, 103

Degeneracy 28
d-d Spectra
  high spin 157
  low spin 162
  tetrahedral complex 160
Descending symmetry 164
Diborane
  MO diagram 95
  SALCs 94
Direct product 121, 127, 140
Dissymmetry 14

Electron-deficient molecules 94
Enantiomer 14

Ferrocene
  Fe AO symmetries 111
  MOs 114
  MO diagram 116
  SALCs 111, 113
  structure 110
Force constant 46

Group
  definition 17
  representations 20
Group frequencies 47, 64

Hole formalism 128
Hydrogen
  linear H_3, MOs 73
  H_2, MOs 72
  triangular H_3, MOs 88
Hypervalency 102

Identity operation 5, 18
Improper Rotation
  definition 7
  axis 8, 14
Infrared Spectroscopy
  basic features 48
  selection rule 48, 62, 130
Inverse operation 18
Inversion
  centre of inversion 7
  definition 7
Inversion Centre 7, 14
Irreducible representation 23

Jahn-Teller theorem 164

Ligand fields 137, 139, 142
Ligand field theory 104

Matrix
  character 25, 35, 41
  definition 19
  chi per unshifted atom 40, 43
Microstates 133, 182

Mirror plane 6, 15
Molecular orbitals
anti-bonding 72, 77
bonding 72, 77
delta, definition 75
ferrocene 114
octahedral complex 103, 104
non-bonding 74
normalisation constant 77
pi, definition 75
sigma, definition 75
rules for filling 73
Mulliken symbols
definition 27
Multiplicity 125, 163

No crossing rule 146
Nodes
definition 73
for cyclic arrays 88, 92, 112
for linear arrays 74
Normalisation constant 77

Octahedral complexes
main group 102
shapes 98
transition metal 103
Orgel diagram 159

Pi-acceptor ligand 106
Pi-donor ligand 105

Point group 8
assignment 12
classification 9
C_1 9
C_i 9
C_n 9
C_{nh} 10
C_{nv} 9
C_s 9
$C_{\infty v}$ 11
cubic 11, 15
D_n 10, 15
D_{nd} 10, 15
D_{nh} 11, 15
D_{4h} 55
D_{5d} 110
$D_{\infty h}$ 11
dihedral 10, 15
infinite 11
I_h 11, 15
T_d 11, 15
O_h 11, 15, 98

Polarisability 50, 51
Polarity 14
Projection operators 60, 100, 173

Raman spectroscopy
basic features 49
polarisibilty 50, 51
selection rule 49, 62, 131
Reduced mass 46
Reducible representation 36
Reduction formula 36, 37, 61
Redundant modes 58
Reflection
definition 6
dihedral plane 6
horizontal plane 6
vertical plane 6
Representation
definition 20
irreducible 23
reducible 36
Rotation
axis, definition 4, 15
definition 4
improper 7
principal axis 5
Rule of mutual exclusion 50
Russell-Saunders coupling 136

SALCs
BH_3 90
B_2H_6 94
$[C_5H_5]^-$ 111
definition 80
H_2 87
H_3 87
H_2O 80
octahedral complexes 99, 104
Symmetry
s-orbitals 82, 90, 102, 111

p-orbitals 81, 90, 97, 111
d-orbitals 101, 103, 111
Selection rules
..band intensities 157
general 123
infrared 48, 49, 130
Laporte 153
parity 153

Raman 49, 50, 131
Spin 126, 156
Spectrochemical series 103
Stretching vectors 39, 56, 62
Sulphur dioxide
vibrational spectrum 39, 50, 52
Symmetry
centre of 7
definition 3
element, definition 4
operation, definition 4
label definition 27

Term symbol 133, 141, 146, 148, 150, 182

Tetrachloroplatinate
mirror planes 6
rule mutual exclusion 50
symmetry elements 55
vibrational modes 56
vibrational spectrum 59
Transformation matrix 19, 20, 34, 41, 56

Vibronic coupling 153

Walsh diagram
NH_3 (C_{3v} / D_{3h}) 92
H_2O (C_{2v} / $D_{\infty h}$) 83
H_3 (D_{3h} / $D_{\infty h}$) 89
Water
Walsh diagram 83
MO diagram 82
SALCs 80

Xenon oxy-fluoride
shape 61
vibrational spectrum 60, 63

Printed in the United States
By Bookmasters